The Nine Elements of a Sustainable Campus

The Nine Elements of a Sustainable Campus

Mitchell Thomashow

The MIT Press
Cambridge, Massachusetts
London, England

MIT Press books may be purchased at special quantity discounts for business or sales promotional use. For information, email special_sales@mitpress.mit.edu.

Set in Sabon by the MIT Press. Printed and bound in the United States of America.

Library of Congress Cataloging-in-Publication Data

Thomashow, Mitchell.
The nine elements of a sustainable campus / Mitchell Thomashow ; afterword by Anthony Cortese.
   pages   cm
Includes bibliographical references and index.
ISBN 978-0-262-02711-3 (hardcover : alk. paper)
1. Campus planning—Environmental aspects—Case studies.   2. Universities and colleges—Environmental aspects—Case studies.   3. Sustainable buildings—Case studies.   4. Community and college—Case studies.   5. Unity College.   I. Title.
LB3223.T47 2014
727.30973—dc23
2013031920

10 9 8 7 6 5 4 3 2 1

After days of anguished speculation about Earth's fate, the Confucian scholar said, we need a heightened sense of human flourishing.

Alison Hawthorne Deming, "Works and Days," in *Rope* (Penguin, 2009)

Contents

Acknowledgments

The development of the "nine elements" concept was an organic, interactive, and reciprocal learning process. Thanks to the staff, the faculty, the students, and the trustees of Unity College for all they taught me about sustainability, leadership, and life. Special thanks to Professors Mick Womersley and Doug Fox, who launched the work before I arrived. Without Jesse Pyles' insight and hard work, we would have accomplished much less.

Thanks to my colleagues at Second Nature, including Anthony Cortese, David Hales, Georges Dyer, Steve Muzzy, Ashka Naik, and Sarah Brylinsky, and at the Sustainable Endowments Institute, especially Mark Orlowski. Working with these people has been immeasurably enriching. I'm thankful for all I've learned from the Steering Committee Leadership of the American College & University Presidents' Climate Commitment, especially Presidents John Anderson, Richard Cook, Michael Crow, Elizabeth Kiss, Judith Ramaley, David Shi, Greg Smith, Timothy White, and Wim Wievel.

Thanks to the many friends and colleagues who have taught me so much about sustainability and leadership: Tedd Benson, Jim Buizer, Michael Catanzaro, Felecia Davis, Harold Glasser, Cary Gaunt, Angela Halfacre, Heather Henriksen, Samuel Kaymen, Stephen Kellert, Amy Knisley, Tom Kelly, Paula Morreale, Gary Nabhan, David Orr, and Diana Van Der Ploeg. I hope I've conveyed even a sliver of their knowledge and experience.

For assistance with the writing process, thanks to the MIT Press reviewers, whose insightful comments helped me sharpen my attention and voice. Clay Morgan's fine synthesis of those comments provided me with a conceptual breakthrough at exactly the right time. Margot Kelley was thorough, and Alison Deming was pithy. They came through when I needed them most. In a previous work I described Chip Blake's editorial assistance as ubiquitous and invisible. This time he was just plain brilliant. Cindy Thomashow was there every minute of every day, providing insight, common sense, and uncommon support.

Introduction: Sustainability Leadership and Higher Education

When I arrived at Unity College in July of 2006 as the new president, I was green in two ways. First, I was naive. I had never been the chief executive of an organization, and I had never worked with undergraduate students or with a board of trustees. I didn't have a sense of the scope of the challenge I had signed on for. Second, I had long been an advocate of environmental studies.

The college, located in Unity, Maine and founded in 1965, was a small one with a great deal of heart and grit, interesting students (560 in all), an experiential approach to learning, and great potential. Its curriculum was rather traditional. It derived its revenue almost exclusively from tuition. Its endowment was minuscule. It was strapped for resources. There was a lot of turnover at the senior administrative level. There was mistrust between the faculty and the administration, and there was direct intervention from the board of trustees. The college was in desperate need of active and engaged leadership.

When the search firm first contacted me about the position, I declined to express interest. For thirty years I had been happily directing graduate programs in environmental studies and developing innovative curricula. Several generations of students had inspired me with their commitment and passion. I had written books and taught courses about bioregionalism, sense of place, the virtues of natural history, global environmental change, and reflective educational practice. I was emotionally attached to the Monadnock region of southwestern New Hampshire. I had a strong community of friends and colleagues, and a seamless integration of place, family, and work. When the search firm contacted me a second time, I again declined to express interest in the presidency of Unity.

After I told my wife, Cindy, about the opportunity, she said I would be foolish not to explore it. Since she is usually right in such matters, I listened carefully to her advice. I decided to apply for the position. I wrote a long letter describing my values and ideals, emphasizing my philosophy of learning and leadership and my thoughts about the future of environmental studies. I didn't discuss my qualifications. I had no idea whether I was qualified to be a college president, even if other people thought I might be.

After two interviews and a visit to the campus, Unity offered me the presidency. I spent a week thinking about why I might take the job. Sustainability issues were becoming increasingly prominent. I was interested in the prospects of sustainability as a way to implement many of my ideas about place, environment, and community. The emerging field of sustainability suggested a solutions-based, participatory way to integrate these ideas. And what better place to experiment with doing so than a college campus? As the president of a college, I imagined, I would be in a position to organize community, learning, and infrastructure, and to integrate classroom study with the practices of everyday life.

Meanwhile, the evidence about climate change was increasingly disturbing. An international scientific consensus was developing. Dozens of panels, consortia, and collaborative research programs were coming to similar conclusions. The message was clear: Unless we dramatically reduce the global carbon footprint, the earth will experience unprecedented, dynamic climatic shifts, and waves of potentially catastrophic environmental change will be unleashed. Governments were stymied by this challenge and seemed to be making little progress in addressing the issue. International protocols were stalled. In public opinion polls, climate change was very low on the list of perceived national priorities. It seemed that the only sector that could potentially make a difference was higher education.

A college or a university is an ideal venue for addressing the global climate crisis. What better place is there to conduct environmental research, to develop curricular approaches, to construct policy mechanisms, to convene multi-sector collaborations, and to implement sustainable solutions? A college or a university has the capacity to engage students, employees, and faculty members in the crucial issues of our times. It can educate people to better understand climate change and to assess its potential impact. It is well equipped to mobilize a shift in awareness, and to demonstrate the relationship among knowledge, commitment, and action. Many campuses were already making strides in developing programs to address the climate challenge. Yet there was much additional potential, and much more work was needed. Colleges and universities, working together, could demonstrate the viability of sustainability while educating a generation of students about the planetary climate crisis. I felt an obligation to contribute to such an effort.

Still, I was hesitant to take on the presidency of Unity College. Why should I give up the security and stability of my seemingly well-integrated life? The day before I was scheduled to sign the contract, I flipped through the position advertisements in the *Chronicle of Higher Education*. Was this really the right job for me? Now that I had opened myself to the prospects for new opportunities, maybe there was a better one! I came across an article, titled "Happily Exhausted," in which a first-year college president wrote that this was the most challenging and difficult job he could imagine, yet he was delighted and inspired by the extraordinary impact he could have. Whether working with the trustees or eating pizza with students in a dormitory, he was thoroughly engaged in meaningful work. Scanning down the page, I noticed that the author was Lawrence Schall, the president of Oglethorpe University in Atlanta. Larry was a friend of mine. For many years we had both attended a basketball camp (for adults) organized by the great professional coach Phil Jackson. I immediately called Larry. After speaking with him, with Cindy, and with several other good friends, I recognized that I had an opportunity to emulate the approach I had always recommended to my students: Take on a new challenge that will test your values and will allow you to make a difference in the world. If I was going to act on my values, it was essential that I take on this challenge, and that I keep my ultimate motivation front and center. University leadership is our last best hope for addressing the global climate challenge, and campus sustainability initiatives are the foundation of that leadership.

When I arrived at Unity, on July 1, 2006, there was no presidential residence. We rented a small cottage in the woods about twenty miles from the campus. Although I was warmly welcomed, I felt somewhat like an immigrant arriving on a distant shore. After thirty years at the same institution (Antioch New England Graduate School, in Keene,

New Hampshire), becoming the president at a new place was an emotional challenge. How long would it be, I wondered, until I felt at home here?

During the first week, the facilities people took me on a tour and expressed their concerns about deferred maintenance. The college's few advocates of sustainability also gave me a tour. They bemoaned lost opportunities, wondering if Unity could ever make sustainability a campus priority. Although I was grateful for the enthusiasm and the high expectations surrounding my arrival, there was much skepticism, too. Many people wondered whether I had the administrative experience, the "technical" sustainability expertise, and the fund-raising capability to be successful. They were correct to have such doubts. I had much to learn about both sustainability and leadership. In addition to the challenge of building sustainability initiatives into the college's mission, I was also going to be challenged to help the college gain confidence in its future. At the same time, I was going to have to find my own voice and leadership style as a college president.

There were many difficulties and setbacks. A conservative, anti-tax board member asked how sustainability would save money for the college and whether it was just another word for environmental regulation. Some faculty members couldn't understand why they should revise the curriculum when doing so would only increase their already heavy workload. The chief financial officer liked the idea of saving money on energy, but raised his eyebrows when I suggested that we consider "green" procurement policies. When I encouraged the senior administrators to incorporate sustainability responsibilities in their job descriptions, they raised many objections. The Director of Advancement told me that there was "no philanthropic potential in sustainability." I am not exaggerating when I say that just about every sustainability initiative I suggested generated at least some doubts. Such resistance to new ideas and change is not uncommon, and in most cases the skepticism demanded that I clarify, simplify, and further explain my motivations.

At every stage, I had to clarify why sustainability is important (especially for colleges and universities), how it is both a way of thinking and a way of life, and why it should be the foundation for how we conceive of service, civic engagement, and the liberal arts. As a personal challenge, I came to realize that sustainability leadership took my ideas about place, bioregionalism, and ecological learning to a new level. How could I integrate these ideas with a comprehensive vision for higher education as embedded in the campus setting? I learned that sustainability leadership is a tangible, everyday, experiential challenge. Having an ideal is very different from implementing it.

Thanks to collaborative work by an extraordinary group of staffers, students, and faculty members, the support of the board of trustees, the growing national sustainability movement, and the receptiveness of the community, we transformed Unity College into a more sustainable campus. We did this on a shoestring budget. We described our approach as "real-time frugal sustainability." Over time, it became clear to me that many aspects of Unity College's success could be framed within the sustainability agenda.

While emphasizing sustainability initiatives, we assessed all the crucial indicators for improving the college's academic and financial standing: selectivity, retention, recruitment, reputation, academic excellence, community service, fiscal stability, fund raising, governance, student life, and campus morale.

Campus Sustainability Leadership

I have always approached leadership using common sense. Build trust, listen well, communicate clearly, serve others, explain your motivations, give voice to opposing points of view, be guided by your core values, and then take action. If you are true to these basic principles, then you will gradually cultivate organizational trust, wisdom, and confidence. This is my leadership philosophy, and it has served me well in most life situations. When circumstances get sticky, complicated, or controversial, these principles become even more important.

But what happens when you add a substantive prefix to your concept of leadership? By doing so, you highlight the values component of your leadership objectives. Specifically, what does the concept of sustainability leadership imply? Sorting this out was the primary conceptual challenge of my presidency. There were five components to this challenge. First, I would have to clarify the meaning of "sustainability." Second, I would have to determine how the principles and behaviors of sustainability might inform my understanding of leadership. Third, I would have to listen carefully to what "sustainability" meant to the folks at Unity College. (They had much to teach me.) I would have to listen equally well to the emerging and very exciting national discussion about sustainability, and to blend all those voices together in ways that would make sense for Unity, but also would allow us to contribute to that national discussion. Fourth, I would have to find the common ground among all these voices.

Our common interest was really *campus* sustainability leadership. How could we implement sustainability practices in this physical place, in our common living and working environment, in a way that builds the strength of our campus community? Here is where my interest in place-based learning and bioregionalism came into play. How would the life practices, the natural and cultural history, and the community values of this place in rural Maine inform our version of sustainability? And how was all this connected to the global climate crisis? This fifth component of the campus sustainability leadership challenge entailed an expanded sense of the word "campus" to encompass the ecology, the economics, and the community life of both the region and the world.

This is the thought process behind the educational philosophy that is central to this book. Before I explain the meaning and details of the nine elements of sustainability, and before I explain my narrative strategy, I'd like to elaborate on the challenges listed above. What is sustainability? How do its principles affect how we think about leadership? Why is it an important concept for higher education? Why is a campus an ideal setting in which to promote and enact a vision of sustainability?

A Sustainability Ethos

I enjoy conversations with people who care deeply about community, the natural world, and living a virtuous life. My travels give me an opportunity to speak with people of different political views living in many different regions of the United States and the world. The message from these conversations is consistent. People want a good life for themselves, their families, and their communities. Despite all the pleasures of affluence and the cultural promises of unlimited wealth and consumerism, most people, when pressed, see through the superficialities of materialism. Not always, and never without the seductions and tensions that accompany the glitzy images of consumer culture. But when you strip away the layers and the distractions and get down to the fundamentals, people want assurances that they will have meaningful work and community, reasonable comfort and security, a clean and safe environment, good health, and opportunities for personal growth. And they want to know that these are enduring qualities.

Ultimately, sustainability is an approach to living and learning that links these qualities to ecological awareness. It seeks to make those connections by calling attention to how personal actions and community practices affect the natural world. Hence, the concept of sustainability simultaneously projects the good life and requires an ecological conscience.

Sustainability is both a promise and a contract. It offers an optimistic vision for an abundant future by investigating ecologically appropriate opportunities, innovations, investments, and discoveries. The promise is that all this can be achieved by thinking more carefully about how we use resources, deepening our awareness of ecological impact, staying open-minded about new technologies, reflecting on the routines and behaviors of daily life, emphasizing educational opportunity, and expanding our ideas about the meaning of community. However, none of this is possible if we don't incorporate and internalize a scientific understanding of earth systems, the biosphere, ecological resilience, and the delicate balance of climate, species, and ecosystems. That is the contract. In pursuing the dream of a sustainable community, we seek to optimize human flourishing in a dynamically changing earth system.

"Ethos" is an appropriate word to link with "sustainability." It refers to the character, the disposition, the morals, and the values that motivate an ideal. It connotes the integration of practical wisdom with virtue and goodness, as linked to community. When we add "sustainability" to "ethos," we convey resilience, conservation, and ecological awareness. "A sustainability ethos" refers to a spirit of creative innovation in support of civic responsibility and ecological resilience.

Sustainability Is a Response to a Planetary Crisis

Why is sustainability necessary? If a sustainability ethos is the moral foundation of your approach to leadership, this question must guide the dialogue. The concept of sustainability is derived from a scientific interpretation of biosphere processes—one that warns of an ecological crisis[1] that may have a dramatic impact on humanity. The twenty-first century's planetary challenge involves three biospheric patterns: species extinctions and threats to biodiversity, rapidly changing oceanic and atmospheric circulations, and altered biogeochemical cycles. These patterns are the planetary stage for human-generated change caused by extensive natural-resource extraction, accelerating consumer demand, and the unequal global distribution of wealth.

Scientific analysis of these changes in earth systems demands the most sophisticated, rigorous, and empirical data-gathering and data-assessment protocols, requiring collaborative, interdisciplinary research. This comprehensive, ongoing research is becoming more precise, definitive, and assertive. Climate change is having substantial effects on all regions of the world and all sectors of the global economy. Whether you review the work of the International Biosphere and Geosphere Programme, that of the Intergovernmental Panel on Climate Change, or that of the United States National Climate Asssessment,[2] there is concrete evidence describing a full-fledged global environmental crisis. Cities and regions around the world must prepare for the dramatic climate scenarios that appear increasingly likely—sea-level change and its effect on coastal cities, catastrophic storms, floods and droughts, wildfires, ocean acidification and its effect on fisheries, human migration from unlivable coastal zones, and various dislocations that may lead to economic deprivation and human suffering. We are entering an era of unprecedented environmental uncertainty.

The sustainability ethos, involving renewable-energy alternatives, reducing carbon footprints, growing organic food, experimenting with resilient and recycled materials, developing low-energy transportation, building community partnerships, incorporating climate scenarios in campus master plans, and promoting innovative technical solutions, is a response to this planetary crisis. Every campus sustainability initiative contributes to this response. Colleges and universities lead the way by implementing these practices on their campuses, working with the larger community to mobilize regional impacts, and building societal awareness of the necessity of a sustainability ethos. Awareness of the planetary crisis and the implementation of sustainable solutions work hand in hand.

Sustainability Is Imperative for Colleges and Universities

A stunning convergence connects the volatile global economy with planetary ecological challenges. On the one hand, individuals, communities, and nations alike are facing extraordinary economic uncertainty. The ubiquitous threat of burdensome debt emerges at a variety of institutional and personal scales, raising concerns about accountability, fiscal priorities, and "conserving" money. The premise is that by overspending we have robbed future generations of their wealth. A proposed solution is to cut budgets, live within our means, stash cash, and invest prudently. Yet the economy will thrive only when there is enough spending and investment to catalyze growth and productivity. The ecological challenge is similar. By "overspending" our resource base, we have "overtaxed" the biosphere, straining the future of human and ecological systems so we can live affluently in the present. And it is today's students—the professionals and the workforce of tomorrow—who are inheriting these challenges. Higher education is at a similar crossroads. It is under enormous pressure to reduce costs, provide the public with value for its investment, and develop measures of accessibility and accountability that balance learning and value. Like governments and municipalities, many campuses are reconsidering their priorities, and giving careful thought to educational and financial resilience.

The campus sustainability movement is an educational response to this suite of planetary economic, ecological, and educational challenges. Its premise is that there is a correspondence between the economy and the biosphere. We have to simultaneously balance our economic and ecological budgets, conserving (saving) our resources and investing them prudently and wisely. Hence, the sustainability ethos asks us to think more carefully about how we live, suggesting that ecological criteria serve as the foundation of economic decisions. The economy should still thrive, produce, and grow, but it must do so in a more ecologically balanced way.

Sustainability leadership requires promoting educational dialogues that explain, interpret, and communicate these challenges. Universities and colleges provide the scientific and intellectual milieu that broadens our understanding of these trends while advancing the solutions for addressing them. The campus must play a prominent role as the educational center for these dialogues. Whether we are describing energy innovations and efficiency measures, campus food-growing systems, "green" construction and materials, or curricular changes meant to improve preparation of the future workforce, colleges and universities can save money, invest in the future, work with the community, promote innovation, and lead the way to a sustainable future.

A campus is an ideal setting for exploring, constructing, and practicing a sustainability ethos. Why? First, colleges and universities take great pride in their legacy. Many of them are among the longest-lasting institutional structures in their communities. In taking their educational mission seriously, they are concerned with inter-generational obligations. It is their responsibility to consider what their students will be doing in future decades. The stature, prestige, and long-term survival of a college or a university depend on this legacy. Second, to ensure this legacy they have endowments, infrastructure obligations, and investment opportunities that are important to the regions they serve. Colleges and universities are always concerned with their financial sustainability, and their thinking about their investments tends to be more long-term than that of most other institutions. Third, in line with their educational function, colleges and universities encourage inquiry, open-mindedness, experimentation, deliberation, service, and intellectual rigor, which are ultimately connected to cultivating virtue and character. Fourth, colleges and universities have the potential to change how people think about the world. They are supposed to ask big, meaningful questions, and they attract people who want to ask them. Fifth, and most important, colleges and universities have an educational responsibility to address the most important issues of the times. The global environmental crisis demands their leadership. Their campuses can generate knowledge, awareness, and solutions that can be used to confront that crisis.

The Nine Elements Emerge

Can an entire campus be organized around a sustainability ethos? Is it possible to fold many of the complicated challenges of university life into this moral compass? Why should any campus want to do this in the first place? Addressing these questions opens a dialogue that informs how we think about sustainability leadership. As the president of Unity College, I used every opportunity to remind the campus why sustainability is important, how it is justified, why it is central to every aspect of campus life, and why it should frame our curriculum. I aspired to generate a campus-wide discussion regarding these issues, promoting the dialogue as an educational opportunity for the campus, and then expanding it to the local and regional community. We considered how a sustainable campus might develop in Unity, Maine, with its history, its habitat, and its bioregion. This discussion became the substrate for all our collaborative processes, including the master plan, strategic visioning, and academic program planning.

As we gained more experience in discussing the implementation of sustainable practices, and as the dialogue permeated the entire campus, we realized that our planning coalesced around three broad concepts: infrastructure, community, and learning.

All campus infrastructures embody a complex interaction between the biosphere and the built environment. Power and energy, sustenance and nutrition, construction and materials—these consumptive processes are the prerequisites of human survival and the source of human ecological impact. *Energy* is more than a way to power our buildings; it is a direct connection between our campus and the biosphere. *Food* is not simply nourishment, but a way to think about soil, landscape, and nutrition. *Materials* are not just the building blocks of our classrooms, but the fruits of processes used in extracting natural resources.

Infrastructure is managed, maintained, and planned by a human community. The community decides how to organize itself, how to sequence its decisions, how to allocate its resources, and how it determines and provides for personal and collective well-being. *Governance* broadens by gathering ideas for sustainable solutions from every campus constituency. *Investment* involves a comprehensive approach for conceiving of wealth, including the far-reaching effects of procurement decisions and endowment policies. *Wellness*, or human flourishing, represents the tangible measure and the ultimate criterion for whether a campus is resilient, prosperous, and vibrant.

A campus is first and foremost a learning community. Campuses transfer knowledge between generations of students, faculty members, and employees. They do so through patterns of instruction, signs and symbols, and works of art and imagination. Ideas about *curriculum* expand to ensure that everyone who has a learning experience on campus is considering how his or her actions affect the earth. A campus serves as a studio for practicing sustainability initiatives. It emphasizes *interpretation* so that every visitor to the campus understands what we are trying to accomplish and is inspired to think more deeply about the meaning and practice of sustainability. *Aesthetics* becomes a campus priority, with art and the humanities used to explore how sustainability initiatives can make the world more meaningful and beautiful.

These are the nine elements of a sustainable campus. On a campus (or, for that matter, at any place where people work, live, and play), these elements inform community decisions. In contemplating their implementation, we tangibly experience how the sustainability ethos informs leadership, and how sustainability can serve as the foundation for the transformation of a campus and a new way of thinking about higher education.

Way Beyond Unity

As I began to speak with other college and university presidents, I came to understand that hundreds of other colleges and universities were facing challenges similar to those that Unity was facing . I had discussions with presidents and senior leaders from every conceivable type of institution. Although their specific circumstances were always unique, patterns could be readily identified. I was challenged to identify those patterns in a way that would provide a versatile approach to sustainability leadership. This would enhance the relevance of our work at Unity College while linking it to the broader challenges of higher education and the national discussion of sustainability. I wanted the Unity College community to understand that what we were doing mattered. Although we were a small college in rural Maine, we were collaborative partners in the national sustainability movement. If we could transform our campus, it might open up the possibility of transforming hundreds of other campuses too.

In order to promote Unity College's reputation, and to test our ideas, I spoke in a variety of venues, including meetings of higher-education leaders and forums on sustainability. I wanted to share our experiences while learning from my peers. Similarly, other institutions were eager to understand how Unity was using sustainability as the basis for a campus transformation. Over time, I realized that the "nine elements" approach was of great interest to other campuses. It seemed to demonstrate scale versatility while addressing leadership challenges and speaking to the needs of multiple campus constituencies. It became clear to me that campus sustainability personnel (regardless of their position) were searching for a comprehensive campus process that could integrate their activities, provide strategic vision, and serve as a flexible guide to action.

More important, I learned that for sustainability practitioners and academics—whether at large land-grant universities, at urban community colleges, or at small private colleges—campus sustainability embodied a deeply rooted suite of values—values stemming from a profound concern about environmental challenges, prompted by a desire to incorporate sustainability initiatives as a way of life. They were interested in building more resilient, enduring, and responsive campus communities. They understood that the campus sustainability movement had the potential to transform higher education, but only if it could reach into every corner of the campus community.

A New Generation of Sustainability Leadership

My daily practice of leadership provided me with the best education I could ever ask for. To better understand the potential of campus sustainability, I had to do a lot of listening and learning, including listening to and learning from students. One of my most memorable experiences came at the Clinton Global Initiative University, an annual nationwide conference supported and moderated by former president Bill Clinton. The CGIU gathers university students who are engaged in innovative service projects on their campuses and in their communities. I was fortunate to attend two of these conferences as one of the hundred college and university presidents who were invited. On the last day of one of these conferences, several thousand students take part in an on-site community service project. Both years I attended, I chose to stick around for that. I thought it would be an ideal way to spend time with students from around the country and listen to their interests and concerns.

I was inspired, humbled, and delighted by the change-making expertise and commitment of these students. They aspired to maximize their impact as community service practitioners. And whether their primary interest was health, poverty, social justice, gender, or environment, they all had a strong interest in sustainability principles. They were interested in soliciting my views on how their home universities could take sustainability leadership more seriously. My response was always to tell them that they should find ways to maximize their political leverage, and to utilize their student voice as a means to do so. These gatherings were especially important to me because they confirmed the vitality of the campus sustainability movement. It became increasingly clear that college and university campuses have a responsibility to empower, educate, and cultivate this new generation of change-making students.

In 2011, I decided to step down from the presidency of Unity College and work as a sustainability consultant. Under the auspices of the sustainability-education organization Second Nature and the American College & University Presidents' Climate Commitment, I began working with a national steering committee of presidents who are leading the sustainability agenda on their campuses. I was invited to consult on sustainability issues in a range of campus environments. I met with business leaders, community planners, energy experts, food growers, administrators, architects, facility workers, scientists, engineers, humanists, and students from every conceivable subject area. I visited campuses of all sizes, public and private, secular and faith-based, in "red" and "blue" states, in all regions of the United States. I met with people of all ages, from diverse cultural and economic backgrounds, and with different political perspectives.

Administratively, a "sustainability center" might be placed with facilities, health and safety, student life, academics, continuing education, the business office, or the president. Many large campuses have schools of sustainability with their own deans and advisory boards. And there are hundreds of new programs and majors, in a variety of interdisciplinary configurations. It is likely, too, that these campuses are forging partnerships with businesses, health-care facilities, utilities, governments, and community groups. The sustainability movement goes way beyond the college campus.

While visiting campuses, I am gratified to discover the emerging technical expertise, the spirit of innovation, and the depth of commitment. Every campus has something unique to offer. Multiple campuses discover similar solutions, locally derived so as to meet unique cultural, geographic, and educational approaches. Sometimes it becomes evident that a campus is developing a promising initiative that is deserving of wider support and dissemination. On many campuses, people are grappling with how to be effective. They are looking for guidance, and they want a meaningful scheme to coordinate their efforts. Ultimately these people see themselves as agents of change, and they wish to work strategically and practically. They aspire to effect change in the realm of the possible, in ways that change behaviors, but to do so with intention and meaning. They believe deeply in the virtues of their sustainability values. They are motivated by an overriding concern about climate change and threats to biodiversity, while aspiring to enhance the prospects for human well-being. This is the new generation of sustainability leadership. With support and encouragement, these people will transform America's campuses.

The Nine Elements in Action

How might campus sustainability efforts best be coordinated as a comprehensive strategy, an engaging collaborative effort, connecting people and ideas from all walks of campus life? How might these initiatives simultaneously strengthen the curriculum and promote the campus as a sustainable learning environment? And how might the campus extend beyond its boundaries and serve as an exemplar for sustainable community practice?

This book presents a philosophical and practical vision for integrating sustainability initiatives in any campus setting. In order for a university to embody a fully sustainable approach to all aspects of its mission, it must conceive of sustainability as a cultural process linked to the habits of everyday life. At its core, sustainability addresses how people live, think, and behave. The sustainability challenge is both subtle and profound, using familiar habits as a basis for transformational change. This book explores the patterns of such change, using the parameters of campus life as a guide. It suggests that all aspects of campus activity are interconnected venues for exploring ideas about sustainability. Decisions, actions, and plans regarding infrastructure, community, and learning, taken together, are the foundation for campus sustainability leadership.

"Taken together" is the operative concept. Universities, regardless of size, can be complex, bureaucratic, and multi-faceted. Even the best-intended sustainability initiatives can get lost in a long list of institutional priorities, or subsumed in a less influential department, or buried by the contingencies of daily campus management An enduring sustainability strategy rises to the forefront of a university's priorities. It permeates all aspects of campus life. It is discussed widely, from the president's office to the trustees, in classrooms and in meetings. Most important, sustainability becomes fundamental to the educational mission of the institution. On some campuses, sustainability progress seems impossibly slow. On others, change comes very quickly. That is why so many factors must be considered simultaneously. And that's why this book seeks to connect them.

Think about all the different ways the nine elements of a sustainable campus are organized. It takes good governance and innovative investment to launch renewable-energy initiatives. Yet without a thoughtful sustainability curriculum and evocative signage or interpretation, the student body or other campus constituencies won't understand why those energy initiatives are necessary. Locally sourced food and materials can reduce energy use while promoting health and wellness. Sustainability initiatives may have an aesthetic outcome as well, contributing to the attractiveness and appeal of the campus. There are dozens of ways one can rearrange these sentences to demonstrate the interconnectedness of the nine elements. Yet connecting these processes is not as simple as it appears. It takes vision, strategy, leadership, and commitment to reiterate that all these processes are the synergistic recipes for enduring campus sustainability.

Chapter 1 introduces the role that energy infrastructure plays in structuring campus sustainability and planning efforts. Chapter 2 emphasizes how gardens and local agriculture can transform a campus's culture and landscape while nurturing diverse community relationships. Chapter 3 suggests how a campus's ecological "footprint" may influence decisions ranging from procurement to design. Chapter 4 explains how sustainability provides a focus for leadership and reviews the incumbent challenges and opportunities. Chapter 5 sifts campus finance through the filter of sustainability, providing an ecological approach to wealth and capital. Chapter 6 explores how human flourishing is central to the ethos of sustainability, connecting personal and ecosystem health to a campus's vitality. Chapter 7 describes why and how sustainability is an adaptive approach to learning. Chapter 8 suggests using an entire campus as an educational venue for linking the built environment to the local ecosystem and the biosphere. Chapter 9 makes the case for imagination as an inspirational fulcrum for sustainability initiatives.

Multiple Sustainability Narratives

I conceive of the nine elements as an unfolding sustainability narrative. I explore my values relating to sense of place, natural history, bioregionalism, and sustainable living in order to explain the roots and narrative arc of my career, and how it informs the philosophy and practice of sustainability leadership. I tell the story of my experiences as a university president in order to ground the narrative in the events, controversies, and tensions of campus life. I recognize that these experiences are merely a node in the network of thousands of similar narratives, taking place at colleges and universities all over the world. To the extent possible, and within the limits of my experience, these narratives are the context for my own. The nine elements are also informed by my work since leaving Unity College. I have had thousands of conversations with sustainability practitioners from campus communities—students, employees, faculty members, senior administrators, trustees, and university presidents. Campus transformation is a collective process, generating multiple sustainability narratives.

Every campus has a unique history and geography. The specific circumstances of a small, relatively young environmental liberal arts college located in rural Maine are very different from those of an urban community college spread across many locations in Southern California. Every campus generates its own challenges, solutions, and processes. But there are also commonalities. I know this is so because when people together from all these different places are brought together they have much to discuss. They learn from the diversity of their experiences.

I open many of the chapters by describing experiences I had as a university president. I tell these stories to provide a context for a broader perspective on sustainability, and also in the hope that these short narratives will enliven the reading experience. Most of the people I have met in the course of my visits to various colleges and universities—students and leaders alike—have wanted to describe their experiences, and have had interesting stories (or case studies) to share. In this book, such stories are springboards for more theoretical discussions. The stories are the common ground. Sustainability narratives are the connective bond between institutions that differ in size, shape, culture, and mission.

Using the Nine Elements

This book takes an integrated approach to campus-wide sustainability challenges—an approach linked to the dynamic complexity of global environmental change, and the local challenges of place, ecological resilience, and leadership. It is meant to stimulate collective creativity, action, and insight. My hope is that readers will use it to connect their experiences to the bigger picture, to understand how any person's work is important, and to see how many people working together can unify an entire campus. The nine elements can be applied as a strategic planning process, and indeed several universities are doing so. In order for a campus sustainability process to thrive, the whole campus must learn and work together.

The nine elements provide a way to explore how a single focused beam of light, or one great initiative, might yield a spectrum of possibilities. One reader may be working on a university's investments and thinking about its long-term financial security. That decision-making process has curricular consequences, too. Another reader may be investigating how a local food initiative, or a community garden, might simultaneously enhance the campus's aesthetic landscape while supporting community agriculture, and might eventually provide the campus with an alternative source of revenue. There are thousands of such reciprocal, mutually enhancing patterns of success.

Throughout the nine elements, I refer to institutional success stories. It is important to understand the scope and the diversity of sustainability initiatives. I do not cover them in depth or assess their effectiveness. Other books do that well.[3] Rather, I point them out to highlight the many paths institutions travel, and, most important, how all these projects are connected, and how colleges and universities can learn from one another.

Each element is presented as the central focus of a chapter that explores ideas about leadership, sustainability, higher education, and the natural world. However, many modern readers prefer to "surf." I have designed the nine elements so you can ride the waves. The book has nearly a hundred sections, and they are listed in the table of contents. My intention is to reiterate the dynamic, emergent, and evocative nature of this material. I genuinely hope that readers will use the nine elements to sustain their respective campuses' unique narratives while always connecting their stories to the bigger picture. The fact that every campus is unique is one of the strengths of North American higher education. There isn't a single handbook, manifesto, or golden path to sustainability utopia. It is hard work, often idiosyncratic, and constantly changing. My greatest aspiration for the nine elements is that that they be suitable for adaptation to a wide variety of campus challenges.

It doesn't matter how much authority one has, how many degrees one has obtained, or how many LEED buildings one has designed. We are working on this together. We succeed and stumble together. We develop expertise while reminding ourselves of the daunting and complex challenges. It is too grandiose to think that we can save the world. But we can do our best to construct thriving communities in our place and time. We are one extraordinary species in a vast assemblage of earth organisms, having evolved through four and a half billion years of planetary change. Throughout this book, I emphasize the earth-system substrate that is the foundation of our sustainability experience and the context for human flourishing.

1

Energy

The Jimmy Carter Solar Panels

Northern New England is a diverse landscape, with vast forests, fields, and wetlands. There are villages, farming communities, and nineteenth-century mill towns in the mountains, in the valleys, and on the seacoast. There are small cities with thriving enterprises (Portland, Portsmouth, Burlington) and older cities that are striving to find appropriate prosperity (Bangor, Manchester, Rutland). There is an interesting mix of natural-resource-extraction industries, small-scale farming, colleges and universities, alternative energy providers, information-based businesses, affluent resorts, and rural poverty. What is most surprising about northern New England is the number of remote communities that are only half a day's drive from New York, Boston, or Montreal.

The Maine town of Unity—the home of Unity College—is off the beaten path. It is situated in a rolling agricultural landscape, away from the prosperous resort towns on the coast and at least an hour's drive from Maine's small cities. Winters are long, cold, and windy.

When I arrived at Unity College in 2006, its program emphasized an environmental approach to the liberal arts. Like many small colleges scattered throughout the United States, it was tuition driven, resource strapped, poorly endowed, and just a few shaky admissions cycles removed from severe austerity. And, similar to other small colleges, it had an idiosyncratic culture and a compelling history.

One story was particularly emblematic of the best of Unity's past. It provided me with a fitting narrative with which to frame the college's approach to sustainability. Thanks to an imaginative staff member and a well-known actress who had a summer house in Maine, the college had acquired the solar panels that had graced the White House during the presidency of Jimmy Carter. They sat on the roof of the restored chicken coop that was now the college's cafeteria. Why had those solar panels, once a bold and encouraging statement about the future of renewable energy, been exiled to a remote college in rural Maine?

I was fortunate to be the president of Unity College between 2006 and 2011, a period that saw a new wave of interest in sustainability programs and projects. It turned out that if we handled the solar panels correctly they could become one of our most important educational assets. To accomplish that, we had to find ways to translate the story of the solar panels to tell the story of the college. Even more important, we could use them to create a gripping sustainability narrative that could link our small college to the biggest energy questions facing the planet.

I would be exaggerating if I told you that we developed a systematic strategy for wrapping the solar panels into our educational mission and story. We didn't. However, I did encourage the senior administrators to use the solar panels to promote the college as they saw fit. It turns out that we received many requests for them. We took most of those requests seriously and various teams of people organized actions, demonstrations, displays, loans, and all manner of deployments. Here are some of the highlights.

Christina Hemauer and Roma Keller, two Swiss filmmakers, made a film, *A Road Not Taken*, telling the history of the Jimmy Carter solar panels.[1] We cooperated fully with their efforts. Various museums were interested in obtaining the panels for displays. We granted them as gifts if the educational merit was fully warranted. Google wanted one for Barack Obama's inauguration. A Chinese solar energy manufacturer wanted one for his solar energy museum in China. He visited campus, we sent him one as a gift, and he invited us to visit his factory and newly designed solar city.

Bill McKibben, an environmental activist and writer and a founder of the grassroots movement 350.org, requested one of the Jimmy Carter solar panels for an interesting and important action. In September of 2010 he helped organize a request urging heads of state around the world to demonstrate their support for renewable energy and climate action. They might symbolically support renewable energy by putting solar panels on their residences. The Obama White House was included in this challenge.

McKibben thought it would be fitting if his group brought a Jimmy Carter solar panel from Unity College to the White House for reinstallation. The full story of this project is told well elsewhere.[2] Let it suffice to say that this was an extraordinary opportunity for the college. Some of our students, employees, and faculty members got involved in the project, and Bill McKibben used the campus to launch his "solar road trip" to Washington via Boston, New York, and Philadelphia, accompanied by three Unity students, two alumni, and our intrepid sustainability coordinator. The learning opportunities were unsurpassed; the publicity was pretty darn good too.

What I wish to emphasize here is the extraordinary educational outreach embedded in how we conceive of energy use. Much of the sustainability movement revolves around how we understand energy, the relationship between energy use and climate, and the virtues of implementing renewable-energy alternatives.

After discussing how we might perceive energy as a naturalist does, I'll explain why energy has been the central focus of sustainability. That will lead to a discussion of energy and the campus. Why is the campus an ideal energy education laboratory? And why is it necessary for campuses to embrace the climate action challenge? I'll also explain how to conceive of energy as a campus landscape feature, or perhaps as a network of distributed intelligence that connects landscapes and regions. I'll conclude the chapter by considering energy as a metaphor and how the energy metaphor can translate to campus energy policy and educational transformation.

Perceiving Energy

In her book *The Energy of Nature*, the mathematical ecologist and naturalist E. C. Pielou presents a compelling exercise in observational ecology. She asks you to consider what you would typically observe during a hike in the countryside—natural features such as the flora and fauna, the landscape, and the weather. She suggests how such a list can be expanded "indefinitely." The next step is to imagine the scene a second time, concentrating on all the "signs of energy." This list includes "twigs and branches swaying in the wind, scudding clouds, flowing water, breaking waves, flying birds and insects, running deer." Then she refers to the sounds you might hear, from the "drumming of rain" to "the hum of insects," reminding us that sound, too, is a form of energy. She evokes images of stormy weather with lightning, wind, and thunder, giving us an additional glimpse into energy's many forms. Pielou reiterates that these are the "attention-getting" forms of energy, then continues to expand and deepen her list including the warmth and brightness of the sun and the growth of plants, and the sun's electromagnetic radiation. She emphasizes the various conversion processes, including that by which plants transform radiant energy into chemical energy.

Pielou encourages the reader to perceive energy as a naturalist would. Her emphasis on the ubiquity of energy is instructive:

Energy in a multitude of forms is as much a part of our surroundings as are tangible things, and it is just as noticeable to anybody who pays attention. In the city, evidence of energy at work—man-made energy—is impossible to avoid: think of the roar of traffic, the bright lights, the construction sites with cranes and concrete mixers, even the din of shopping mall music. But energy is as abundant in the tranquil countryside as it in the city, since all energy has its ultimate origin in natural sources exactly as material substances do. Imagining otherwise is like a city child's not believing that milk comes from cows because it so obviously comes from cartons.[3]

That last comment is pertinent. You can't possibly understand the complexity of energy, life, and the earth system if you ignore its presence. And that is something we all tend to do. We mainly think about how much energy costs. We are most reminded of its necessity when it is absent or scarce. We fear energy scarcity because of its potential economic impact, or how it might reduce our comfort. The most dynamic way to appreciate the necessity of energy is to experience a blackout, or to be without a form of consumable energy when you need it most. It is similar to being near the ocean on a hot day when you have no water to drink. Energy is ubiquitous. The challenge is always in its conversion.

By perceiving energy as a naturalist would, one can connect all ecological processes to energy relationships. At least one can know what questions to ask. As I write this section, I am aware of the various forms of energy that surround me. The warm September sun shines through my window, heating my room and my desk. The loud hum of insects signals the end of summer. An airplane, burning fossil fuels as it cuts through the atmosphere, projects sound waves into my study. My laptop computer is powered by my house's electrical system, connected via the grid to either the Seabrook nuclear power plant in New Hampshire or a source of hydroelectric power in Canada. The connections are vast and intricate. Each of these processes contributes to the earth's energy budget, climate emissions, biosphere processes, information and communication pathways, and even knowledge creation. Yet it is unlikely that any of this would be on my mind if I weren't making a concerted effort to call attention to it.

John Muir loved to climb trees during snowstorms because he understood that the direct experience of energy in nature is inspiring. Extraordinary weather events—severe thunderstorms, tornadoes, and blizzards are humbling and daunting, reminding us that the earth is a complex energy system, powered by the sun in ways that we still don't comprehend entirely. Then there are the shocks and movements of the earth itself—earthquakes, tsunamis, landslides, floods, and all manner of geological processes. If we similarly expand our view to conceive energy in time as well as space, we realize that fossil fuels, nuclear power, and hydroelectric power link life processes to the biosphere.

If we are too busy to summon daily observations of energy, perhaps we can call attention to the processes that most affect our daily lives. There is nothing more simultaneously tangible and mysterious than observing the energy-conversion process. We have the illusion of mastery by virtue of the ever-present switch. That's how we "power up." We know how to turn on the lights, put the key in the ignition switch, get the stove running for a cup of tea, and find the remote control for the television. We learn how to control energy-conversion processes from the first day we turn a switch. But do we really know what we have done? And why should we?

A good way to think about how we perceive energy is to consider movement and transportation. While waiting to retrieve my luggage from a luggage carousel in an airport, I sometimes think about the hundreds of thousands of pieces of luggage moving around the world. Is our species characterized by its incessant desire to move things from one place to another? How much energy does this take, and where does that energy come from?

"Energy" refers to the ability to do work, which involves transforming (converting) of matter to produce heat and electricity. The point of sustainable-energy practices is to maximize the efficiency of those processes so as to minimize unwanted by-products. The underlying idea is that we need an energy algorithm that enables us to heat and cool our buildings, move people and their goods from one place to another, and power our machines without simultaneously altering the biosphere. Perceiving energy requires that we combine a naturalist's view with an understanding of how to generate power for human use.

Sustainability can serve as the curricular foundation for a deeper recognition of the ecological necessity of energy. Let's broaden our comprehension of energy beyond cost efficiency and carbon counting to an understanding of life itself. The conversation about sustainable energy goes much deeper than it first appears.

Energy Is Paramount

Since the early 1970s, energy policy has informed sustainability initiatives. It is the issue that the American public is most likely to link to environmental concerns. During the 2012 American presidential campaign, the party platforms discussed energy policy, and made some fleeting reference to climate, but mainly avoided any discussion of other serious environmental matters. You have to search long and hard to find any campaign debates about threats to biodiversity, species extinction, ocean acidification, or altered biogeochemical cycles. The reason for this is that energy is tangibly connected to politics, economics, and national security. Consider just some of the ways in which energy issues enter public awareness. The "energy crisis" of the 1970s (which connected long lines at the gas pumps with Middle East turmoil), the steadily rising costs of gasoline and home heating fuel, the broad public awareness and controversies regarding the relationship between fossil fuels and climate change, the cries of "Drill, baby, drill," and the proposals for supporting a "green" economy all significantly affect how we think about energy. Let us not forget the deep concern generated by the Three Mile Island (1979), Chernobyl (1986), *Exxon Valdez* (1989), Deepwater Horizon (2010), and Fukushima (2011) incidents.

Disasters capture our attention for many reasons, not the least of which is the perception (or prospect) of an energy technology system that is temporarily out of control, threatening rampant toxicity, waste, or the danger of explosion. This is not the place to discuss the relative risks, merits, and efficiencies of various energy technologies. Rather I wish to suggest that the possibility of energy disasters is an important factor in understanding why energy continuously reenters public awareness.[4] Renewable energy is a policy response to this "control" issue, emphasizing that smaller-scale, decentralized, less toxic, less wasteful energy production approaches are an alternative path toward safety and security.

Current ideas regarding sustainability originated in the late 1960s and the early 1970s.[5] The "soft energy" path was launched by Amory Lovins and dozens of his colleagues, who advocated achieving energy efficiency by combining rigorous conservation and retrofitting with passive solar technologies, low-cost traditional building techniques, and decentralized power sources while researching how to maximize the effectiveness of renewable-energy technologies. These ideas were implemented as solutions to the challenges of energy dependency on foreign oil, the proliferation of nuclear power plants, and the centralization of energy production and distribution. This emphasis had a brief flowering of influence, culminating in presidential support in the form of the Jimmy Carter solar panels. Dozens of campuses, inspired by the urging of students and faculty members, implemented "soft energy" solutions. This period was also characterized by the first wave of environmental studies and energy education programs on college campuses.

I vividly recall watching President Carter, in the midst of the energy crisis of the late 1970s, make an energy speech to the nation on television. Wearing a sweater, he implored Americans to turn down their thermostats, suggesting an era of sacrifice as a bridge to a promising renewable-energy future. In contrast, during the 1980 presidential election Ronald Reagan promised an era of unparalleled affluence, impugning the prospect of sacrifice and questioning any attention to energy conservation or renewable sources. Upon his arrival in the White House, the solar hot water panels were promptly removed, eventually to wind up at Unity College. This contrast metaphorically anticipates four decades of energy controversy.

Stephen Schneider's 1989 book *Global Warming: Are We Entering the Greenhouse Century?* prompted a new portfolio of environmental concerns directly related to energy use.[6] The controversy surrounding fossil fuels was now heightened by the greenhouse hypothesis, which suggested that carbon emissions from fossil fuels would generate irreversible and unpredictable climate change. This called new attention to the necessity of developing renewable energy sources; it also stimulated criticism of climate scientists. (These controversies, ongoing still, are comprehensively described elsewhere. The accounts make compelling reading for anyone who is interested in the public communication of science.[7]) Climate issues became the transcending environmental concern, trumping biodiversity, species extinction, and environmental pollution in terms of public attention and awareness.

This brief history explains why energy became the most prominent environmental issue. You may not know enough about the natural world to observe an endangered species, but you can turn down your thermostat, retrofit your home, and ride a bicycle to work. Moreover, the economic context of energy is often directly apparent. By adopting energy-conservation measures you simultaneously save money and save the planet. No more reminder is needed than the little notes in hotel bathrooms that inspire you to save the planet by reusing your towels. I do not wish to make light of this helpful but superficial request, only to point out that public awareness of the energy issue became the basis for many popular interpretations of sustainability.

In 2005, as the evidence of climate change became more conclusive, and it became clear that international protocols for addressing the issue were frustratingly slow, the sustainability movement took on a new urgency. A new wave of sustainability initiatives swept through colleges and universities, businesses, towns, and cities, creating new opportunities for communities and consumers. In higher education one of the most significant initiatives was the American College & University Presidents' Climate Commitment (ACUPCC).

Climate Action Planning

The philosophy behind the ACUPCC is that if you can mobilize college and university presidents to commit their campuses to climate action planning, you can accomplish four objectives in a single stroke: (1) reduce the campus's carbon emissions and address infrastructure challenges, (2) build curricular approaches throughout the university, (3) build the collective voice of higher-education leadership in support of sustainability, and (4) influence all campus constituencies, from students to the board of trustees.

Here is how it works: An ACUPCC signatory pledges to achieve a zero-carbon campus within a specified period of time, depending on a variety of campus contingencies. Carbon neutrality is essentially an accounting procedure. A campus balances the carbon it releases with sequesters and offsets. This is a peer-reviewed process with a rigorous but flexible protocol requiring greenhouse-gas inventories, a climate action plan approved by the college's leadership, and periodic progress reports. The climate action plan becomes a comprehensive template that integrates sustainability, governance, and financial initiatives to maximize the prospects for climate neutrality.

When Dr. Anthony Cortese, the founder of this project, initially contacted a dozen prospective charter signatories, he had no idea that within five years nearly 700 institutions would join the effort, forming a powerful network of campus sustainability leadership authorized and organized at the presidential level. Although some campuses with prominent sustainability policies and programs avoided the commitment, it was clear that momentum from it influenced their efforts. In five years, in coordination with other data-oriented protocols, including the Association for the Advancement for Sustainability in Higher Education's STARS (Sustainability Tracking and Rating System) program, higher education took a major leadership role in advancing sustainability awareness and accomplishments.

Although the ACUPCC integrates campus sustainability initiatives, its central focus is climate action planning. Hence the program recommends establishing a renewable-energy portfolio, emphasizing energy efficiency and conservation, promoting campus-wide education efforts and constructing feasible financial scenarios. Again, energy is paramount in organizing campus sustainability efforts.

Climate action planning is of particular significance because it provides a way to link carbon emissions and energy pricing—to link a biogeochemical cycle (that of carbon) with an economic approach. I will address this in more detail in chapter 5. Here I wish to emphasize that climate action planning, under the auspices of the ACUPCC and other measurement systems, provides a systematic way to organize campus sustainability efforts. It does so at a scale (the campus) that allows for short-term and long-term planning, incorporates economics and sustainability, provides measures of accountability and progress, and involves all campus constituencies. A campus is an ideal scale for experimenting with renewable energy and reporting on the results. It makes good sense that energy should be a central focus. Let's further investigate the relationship among energy, climate action planning, and scale.

Energy and the Campus

E. F. Schumacher's 1973 book *Small Is Beautiful* influenced the formation of the sustainability ethos. Schumacher introduced the mutually
reinforcing ideas of appropriate scale and intermediate technology. Essentially, Schumacher suggested that scale ought to be a prominent value
in all planning choices, but especially in decision making regarding the
use of technology. Further, he advocated a blend of traditional and advanced technologies, depending on the context. The subtitle of the book,
"Economics as if People Mattered," meant that economics and planning
should be pursued on a "human scale." Of course, there are many ways
to interpret that. Mainly it was conceived as an emphasis on quality of
life. Schumacher's view influenced the idea of bioregionalism, the concept
that ecological and geographical features should determine political and
economic jurisdictions.[8] It had a major influence on the emerging idea
of "soft energy." In the 1970s, solar energy, wind turbines, wood chips,
and conservation initiatives were labeled "appropriate technology." The
emphasis was on human access and participation, decentralization, and
local control. Forty years later, these criteria are deeply ingrained in the
sustainability ethos.

Consider the university campus as an ideally scaled location to implement these ideas. Almost every campus, whether urban, suburban, or
rural, has an important economic and environmental impact on the surrounding region. Most campus constituencies highly value "human-scale"
decision making as reflected in governance (participation and planning),
community partnerships, and the quality of life. Colleges and universities are at the forefront of sustainability ideas not only because they are
educational centers of intellectual energy and activism, but also because
they are the ideal scale for implementing sustainability initiatives. In particular, decisions and actions regarding energy infrastructure ought to be
informed by scale. Campus energy policy is a means to incorporate planning, participation, behavior, frugality, leadership, and legacy—criteria
that are relevant to "human-scale" decision making. When such decision
making occurs, multiple campus constituencies can work collaboratively
to reduce carbon emissions, reduce expenditures, and construct a comprehensive energy strategy. Taken together, these factors also represent an
excellent opportunity for collaborative learning.

Climate action planning embodies the "human-scale" approach. First, campuses initiate a collaborative, comprehensive assessment of energy inputs and outputs, including a greenhouse-gas inventory, and compare financial costs (and savings) with carbon emissions (and reductions).[9] Most instructively, they map the flow of energy into campus, further assess campus energy production, and trace emissions. This is best done with both numbers and visual illustrations. Every campus should have a map of energy production and consumption depicting inputs and outputs.[10] Such a map will amply depict emissions of greenhouse gases (GHGs). The second step is to develop a process for reducing GHGs as the foundation for a more comprehensive campus energy plan. During this process, the campus considers energy-efficiency scenarios, building retrofits, and renewable sources of energy production. Depending on the "scale" of the operation, various ecological and geographical factors, the availability of energy resources, and the importance of local control, the campus will construct its energy future. An emphasis on climate neutrality will take all these factors into account.

The planning process should be highly participatory, including as many campus constituencies as it can. The campus energy plan will have a major effect on the future of a school's finances, architecture, and infrastructure, so the school's leaders, its trustees, and its alumni will want to have a voice. So will the ultimate users, including employees, students, faculty members, and community partners. In general, the more inclusive and participatory the energy plan, the more people will be invested in its outcome. As more people become invested in the outcome, the prospects for conservation awareness and behavior change will be enhanced.

Energy consumption can be reduced most effectively at the scale of human behavior. People, not buildings or cars, are the ultimate users of energy. We choose how we use energy. The full measure of our use is reflected in our consumption habits. Yet energy-use behaviors are more likely to change when there are suitable infrastructures affecting and influencing behavior.[11] It is much easier to activate energy-saving behaviors when it is clear how to do so. If students live in an energy-efficient residence hall or take classes in a building powered by renewable energy, and if those buildings include interpretive designs that demonstrate how to participate in the use of the building, it will be much easier to reduce the campus's energy use. Such behavior change is intrinsic to campus energy education.

There are now dozens of such "energy behavior" approaches on campuses, ranging from bicycle-sharing programs to conservation competitions. One major challenge is to incorporate "hands-on" human-scale behavior into a participatory campus energy planning process; another is to emphasize transparency of energy production, energy distribution, and energy consumption when designing buildings and programs.

A human-scale approach emphasizes frugality, the idea of doing more with less. This approach is more likely to surface during challenging economic times; however, frugality is also a philosophy of life, and it is surely intrinsic to the sustainability ethos. We customarily associate frugality with saving money, or at least with trying to get the maximum value out of every dollar. The same approach yields great educational benefits when linked to how energy is used. Campuses are compelled to emphasize frugality in cutting costs, in being mindful about tuition, and in maximizing the educational value of the tuition dollar. Indeed, this is the rationale that some campuses employ when trying to reduce energy costs: If we can reduce expenditures on energy, we will not have to increase tuition as much as we otherwise would. Yet frugality ought to go much deeper. It should be linked to every natural-resource-extraction process on the campus, especially energy use. The best way to demonstrate frugality at a human scale is through tangible energy-conservation measures.

The importance of leadership to human-scale decision making can't be overemphasized. Campus leadership sets a standard for cultivating relationships, accumulating social capital, and building confidence in comprehensive energy planning. The human scale of leadership is displayed in day-to-day relationships, behaviors, and actions. If the president and all other campus leaders make concerted efforts to live sustainably, to support energy innovations, and to engage the campus in collaborative behavior change, the campus will make huge strides. Leadership works when it touches people emotionally as well as conceptually. Leadership is most effective when it can explain the meaning behind and the reasons for its actions. Leadership must always consider the human scale of its policies.

Leadership is well served by referring to legacy—by asking "How do our current actions build on the best ideas of our predecessors and help us to create a better life (or campus) for future generations?" Energy planning, especially in response to the planetary climate challenge, is necessarily forward looking. Though it may be difficult to fully grasp how climate change will affect a community, it is always possible to invoke an inter-generational mandate. Legacy is best championed at a human scale. What is more tangible than keeping a young child in mind as you make plans for the future? A college campus has a unique opportunity to invoke human-scale, intergenerational thinking as a prerequisite to comprehensive energy planning. The questions to ask are these: "What will our campus look like in the future?" "Who will be living, working, and studying here?" "How will our current actions and plans provide them with the best possible learning environment?"

The Campus Energy Landscape

Energy production has a significant imprint on an ecological landscape. When flying over West Virginia, one can observe the dramatic impact of coal mining. Wind energy farms, offshore oil rigs, oil refineries, nuclear power plants, and solar panel arrays leave profound visual impressions. Yet linking one's use of energy at any given moment to its direct source is perceptually challenging. As I noted earlier in this chapter, while writing these words on my laptop computer I didn't conjure up images of the Seabrook nuclear power plant or of a hydroelectric facility in Quebec. Yet when I lived in Unity House, I often considered the relationship between my energy behaviors and actions. Sustainability initiatives typically intend to illustrate these perceptual connections.

At Unity College, we built the first—and as of this writing the only—"LEED platinum" college president's residence in North America. (LEED, standing for Leadership in Energy and Environmental Design, is a standard "green construction" protocol.) The 1,900-square-foot Unity House, built in partnership with Bensonwood Homes, was a prototype design featuring a hybrid of active and passive solar heating technologies, state-of-the-art insulation engineering, zero carbon emissions, modular construction, and low waste. It was planned to be both public and private, suitable as a dwelling for a small family and also available for community functions. Because the budget for the project wasn't large, and because we desired to demonstrate its affordability, we kept the costs down.[12]

We were well aware of the philanthropic and pedagogical potential of Unity House. It was important that the college's president model the sustainability behaviors that he expected the entire campus to embrace. We assumed that Unity House would attract hundreds of visitors who would be interested in the house for purposes of sustainability education. And we invited potential donors, explaining how Unity House was the first of a series of Unity College projects that would enable us to reach carbon neutrality. We aspired to establish the campus as an exemplar of energy education.

Over the course of a few years, Unity College's grounds were covered with new sustainability projects. Essentially, the campus received a sustainability makeover. This was achieved through strategic philanthropy (lots of small grants), student and faculty initiatives, and a spirit of experimentation. Anyone who approaches the campus immediately notices solar panels, small wind turbines, a wood-chip boiler, and other examples of renewable energy, all seamlessly integrated with the campus. The college became a renewable-energy landscape. These features were enhanced with widespread vegetable and wildflower gardens (see chapter 2) and artistic embellishments (see chapter 9). In effect, these energy installations became symbols of sustainability as well as symbols of institutional change. The visual impression of the landscape was reinforced with photographs of these projects on almost all of the college's promotional material. Yes, we "branded" the college as a place where sustainability was embedded in the physical and learning landscape. Yet "branding" doesn't convey the profound value shifts that accompanied these changes.

Renewable-energy installations dramatically change the physical landscape of a campus. And those landscape changes convey a deeper meaning about sustainability values. They convey an image of campus that is forward looking, research oriented, and engaged in thinking about its future. The evidence for this is easily obtained by scanning the promotional materials of colleges that are actively engaged in climate action planning and sustainability. Visit their campuses and read their literature. Among the dozens of examples, some of the following stand out. The campus of Arizona State University has dozens of solar installations. The University of Minnesota's Morris campus prominently features wind turbines. The Georgia Institute of Technology has a solar array on the roof of its recreation center.

Renewable-energy installations change how a campus community perceives energy. First, they establish the campus as an active producer of renewable energy, reducing climate emissions and gaining more control of its energy future. Second, constructing such facilities on a campus, or in close proximity to it, builds energy production and consumption into the daily behaviors of students, employees, and faculty members. Third, it develops energy partnerships and networks with other institutions and organizations—the campus becomes a decentralized regional energy hub. Fourth, it establishes the campus as an energy leader in the community. Fifth, the visibility of these efforts evokes inquiry, emulation, and response. When an educational institution demonstrates leadership, people take notice and consider how they can improve on what they are seeing or how they can modify it for their own purposes. These projects spark innovation.

Renewable-Energy Networks

Campus renewable-energy installations constitute a national campus energy network. If you consider all the solar arrays, wind turbines, geothermal projects, biomass production facilities, conservation retrofits, energy-efficient designs, and other research initiatives on college and university campuses, you realize the magnitude of these comprehensive changes. Of course, the challenge of the required transformation to renewables is still daunting, yet this revitalized energy landscape is hopeful and inspiring. What would happen if all these efforts were coordinated into an intelligent, distributed energy grid?

In his book *Hot, Flat, and Crowded*, Thomas Friedman constructs a scenario depicting what it is like to "live inside a real green revolution" in the imaginary year 20 E.C.E. (by which he means the Energy-Climate Era). Consider Friedman's proposal for an energy Internet "in which every device—from light switches to air conditioners, to basement boilers, to car batteries, power lines and power stations—incorporated microchips that could inform your utility either directly or through an SBB (Smart Black Box) of the energy level at which it was operating, take instructions from you or your utility as to when it should operate and at what level of power, and tell your utility when it wanted to produce or sell electricity." Friedman is suggesting the implementation of a "smart grid" that links networked communication with energy distribution. This can easily be integrated with contemporary technology. What we lack is sufficient infrastructure, and that lack is due mainly to an outmoded political economy of energy.[13] In *Reinventing Fire*, Amory Lovins and colleagues elaborate on the benefits of smart-grid energy networks:

Integrating information technology with electricity enables enhanced grid intelligence and price transparency, making every part of the system cheaper to run and better coordinated. An information-rich electricity system also enormously expands the range of offerings from which traditional and new service providers can assemble new value bundles for every taste and purse. And by enabling smaller, more granular, shorter-lead-time projects, this shift of electricity sources and scale can help utilities and capital markets manage their increasingly worrisome asset-related financial risks.[14]

Colleges and universities have the potential to expand the concept of the energy network. There are several ways this might emerge. Individual campuses can work within their own jurisdictions to establish local smart energy grids connecting facilities and functions. Or they can work within already established higher-education consortia to combine efforts on the state or the regional level. They can implement cross-sector partnerships with businesses, health-care facilities, museums, and other organizations, working with local utilities to create regional smart grids. Or, most ambitious, they can construct a national Energy Internet for colleges and universities. With effective metering, climate action planning, comprehensive greenhouse-gas inventories, and innovative financing mechanisms, colleges and universities are building the capacities for gathering and sharing information that will make it possible to take this next step.

Developing shared energy infrastructures entails many political, economic, and social challenges, but within five years nearly 700 colleges and universities agreed to work together to promote climate neutrality on their campuses. What is to prevent them from implementing a system of collaborative energy networks? I hope that Friedman's vision will one day come to pass. And the renewable-energy landscape will include more than just solar arrays and windmills. There will also be smart buildings, grids, and networked systems that connect sustainable-energy initiatives in an integrated framework.

Entropy and Infrastructure

As I have already mentioned, during my first week as president at Unity College the buildings and facilities people took me on a tour of the campus and showed me all the rot and other deterioration. In those early days

I was the recipient of many supplications. Everyone perceived his or her issue as the most urgent one. Still, I was unprepared for the sheer scope of the infrastructure challenge. Shortly thereafter I had my first series of meetings with the chief financial officer. Dealing with finances and dealing with buildings were the two aspects of the college presidency that were farthest from my previous experience. My challenge was to keep from being unduly influenced by people who claimed to have more expertise and experience than I. My advantage was that I looked at everything with a beginner's eyes. I was able to ask many of the basic questions that one would expect someone so inexperienced in such matters to ask.

When I first went over the "capital assets" section of the college's budget, I was very confused by the depreciation column.[15] I wondered about the underlying assumptions regarding depreciation formulas. How, I wondered, did we arrive at the standard depreciation measure? I was told that these were typical accounting procedures. You balance the expected depreciation with additional infrastructure investment. After several rounds of questions I understood the economic assumptions behind the formulas, but I never completely grasped whether there was any common sense behind it. Why? The formulas were based entirely on abstract economic approximations, and were never adjusted. They weren't informed by ecological and/or energy relationships. If a building is made of resilient, ecologically sound materials, do we consider its depreciation differently than we would otherwise? What about the relationship between depreciation of facilities and depreciation of the ecosystem? I kept most of these questions to myself, at least initially, but they further enhanced my resolve to bring ecological cost accounting into all the standard financial formulas.

I found that the concept of entropy offered a much better way to conceptualize depreciation and its physical manifestation, deferred maintenance. The Second Law of Thermodynamics was my way of connecting economic depreciation to energetics and ecological systems. Infrastructure requires the integrated organization of energy and materials. Over time, any building will slowly dissipate and reenter the biosphere. By thinking about energy and materials this way, I was better able to integrate sustainability concepts with the rot in our buildings. Every standard measure of financial depreciation I encountered failed to connect infrastructure to entropy in an ecological way. I had to interpret these accounting measures so as to incorporate sustainability criteria.

Almost every college president, CFO, building and facilities engineer, or sustainability officer I encounter spends an extraordinary amount of time dealing with the relationship between entropy and infrastructure (or, stated differently, sustainability and deferred maintenance). Unity College didn't have enough buildings for its academic needs. When I visit a college that seems rich in buildings and space, the president often complains about the problem of deferred maintenance, wondering where the funding to fix the buildings will come from. The consensus is that putting up a new building is much easier than fixing an old one, and that the energy costs of a new building will be lower.

The ultimate campus energy-conservation challenge is to make old buildings more energy efficient—for example, to replace old boiler systems with renewable-energy installations, and to insulate buildings that were constructed decades ago. How can a campus with old buildings, and with heating systems that burn fossil fuels, be expected to achieve carbon neutrality? This is the most important question I consistently hear. It is one of the major frustrations that sustainability professionals (at any leadership level) are most likely to express. The entropy/infrastructure challenge seems utterly daunting. Linking it to broader ecological and energy concepts may be a wonderful pedagogical gambit, but it doesn't get us any closer to solving our problems unless it can be tied to cost savings.

Yet Amory Lovins and his team of researchers at the Rocky Mountain Institute have amply demonstrated that combining energy conservation and timely retrofitting is the most cost-efficient way to address the entropy/infrastructure challenge: "One of the most powerful conclusions about the cost-effectiveness of energy efficiency . . . is how it compares to the cost of the energy it displaces. . . . Whether compared to electricity or natural gas, energy efficiency in buildings typically costs less than half as much as the energy it saves."[16] Further, they suggest that the timing of upgrades and retrofits is crucial: "Every building undergoes cyclic changes throughout its life. Major systems like facades and mechanical systems wear out, leases expire, sales and refinancings and market changes occur. As a result, astutely timed energy retrofits can piggyback on changes being made anyway, greatly reducing capital cost."[17]

This approach to energy and buildings is becoming more routine for colleges and universities. There are now hundreds of conservation and retrofitting innovations that blend efficiency, innovation, and strategic financing. Further, Lovins suggests that these approaches ought to be co-ordinated nationally. He lists six main imperatives:

Attack buildings' inefficiencies with transdisciplinary insight and entrepreneurship.
Make energy use more transparent.
Provide easy-to-access financing, priced commensurate with energy efficiency's exceptionally low risk.
Train and educate a high-quality workforce.
Upgrade to next-generation building efficiency policies and align utility incentives.
Begin overhauling how building design is done, taught, and built.[18]

I highly recommend that anyone involved in campus energy planning read *Reinventing Fire*, in which Lovins and his colleagues clearly integrate tangible, hands-on, cost-effective consumer and industry applications with systematic policy objectives. It is particularly gratifying that universities, through the collective voice of the ACUPCC and the emergence of so-called green revolving funds (see chapter 5), have ways to share the best practices and to make the case for regional and national campus energy planning. There are no longer any excuses. With the right leadership, any campus can make strides in dealing with the entropy/infrastructure conundrum. The question, really, is "How far are you willing to go?"

Energy as Metaphor, Metaphor as Policy

Two expressions that are often heard when colleges and universities address the future of higher education are "curricular renewal" and "curricular transformation." There is a compelling underlying assumption behind these terms. To be effective and pertinent, teaching, learning, and career preparation must continuously adapt to change. Struggling colleges and universities use these expressions—"curricular renewal" and "curricular transformation"—to convey flexibility. Elite and flagship institutions promote themselves as innovative and creative. This language attracts students, staffers, and faculty members, and is used as the basis for many fund-raising campaigns. Yet many colleges and universities (even the smaller ones) have entrenched philosophies, bureaucratic impediments, or powerful constituencies that inhibit change. Suggesting the importance of renewal or transformation is much easier than implementing such change. (I cover this challenge in much more detail in chapters 4 and 7.)

What does any of this have to do with energy? While reading *Reinventing Fire* (an outstanding practical guide for any organization that wishes to revitalize its approach to energy), I was struck by a metaphorical parallel. In a chapter titled "Electricity: Powering Prosperity," Lovins and his collaborators propose four electricity scenarios as a means for "imagining" America's next electricity system, with the ultimate goal of producing 80 percent from renewable energy sources. These scenarios are evaluated using five criteria: technical feasibility, affordability, reliability, environmental responsibility, and public acceptability. Here are the four cases in brief:

• *Maintain* assumes that the future system looks largely like today's system, in both demand and supply mix.
• *Migrate* assumes that the anticipation of legislation to reduce greenhouse gas emissions drives a switch from conventional fossil-fueled generation to more nuclear power and to new coal plants equipped with carbon capture and sequestration (CCS).
• *Renew* examines how renewables like solar, wind, geothermal, biomass, and hydro can provide 80% of U.S. Electricity.
• *Transform* picks up where *Renew* leaves off. It is powered by resources of varied scale but includes more distributed generators such as rooftop solar, CHP, fuel cells, and small-scale wind.[19]

An interesting convergence emerges. These policies for the future of U.S. energy are also relevant for college and university campuses. Every campus has choices it can make in planning its energy future. You can't have a climate action plan without specifying in detail how you will reduce carbon emissions and meet energy demands. How the campus chooses to pursue its energy future will tell you a great deal about its leadership, its values, and ultimately its curriculum. My experience suggests that these metaphors—maintain, migrate, renew, and transform—will simultaneously reveal how the campus plans both its energy future and its curricular future. There is not only a metaphorical convergence but also a reciprocal innovative process. As you transform your campus's energy future, you develop mutually reinforcing infrastructure, curriculum, and investment processes. Indeed, it is unlikely that you can transform your energy future without innovative financing mechanisms, resilient community partnerships, progressive governance, and flexible curricular integration.

When campuses are stuck in energy scenarios that resemble "maintain" or "migrate," it is typically because they can't extricate themselves from some combination of burdensome infrastructure, restricted finances, limited leadership, change averse boards, and staid academic programming. These situations breed excuses, inaction, frustration, and cynicism.

We have all experienced these impediments to change. What is hopeful is the sheer number of colleges and universities that are overcoming these impediments. They are actively engaged in constructing truly transformational approaches to energy, investment, governance, and curriculum. We now have good examples of "best practices" that are embracing the "renew" and "transform" scenarios, and many of these cases provide good examples of how to overcome the standard obstacles. The variety of the institutions involved—campuses of every conceivable kind, some small, some large, some private, some public, some resource strapped, some prodigiously endowed—is noteworthy. There is profound interest in the "renew" and "transform" metaphors, and there are case studies[20] to help show the way.

The Jimmy Carter Solar Panels were a useful symbol for Unity College because they represented the college's desire to keep the dream of renewable energy alive and to call attention to the energy landscape of our campus. Although they lay dormant on the roof of the cafeteria for many years, they were a reminder of what the college stood for and what it hoped to accomplish. Eventually the moment arrived when the spirit of campus (dare I say "campus energy?"), the growth of the sustainability movement, and the panels' emerging historical interest allowed us to utilize them as a symbol of campus transformation. Unity College is now an interesting energy landscape, and its curriculum emphasizes sustainability science. Its energy future and its curricular destiny are intertwined. Though just a small school in a remote corner of northern New England, Unity College is a node in a nationwide campus renewable-energy network that network is still in its early days.

2
Food

Three Campus Meals

For a cold place with a relatively short growing season, northern New England has a surprisingly robust network of local agriculture and organic farms. The Maine town of Unity is an active participant in that network. The Maine Organic Farm and Growers Association holds its annual Common Ground fair in Unity.[1] It is a remarkable gathering of organic agriculturalists, craftspeople, vendors of renewable energy, and sustainability advocates. Unity College's proximity to the fairgrounds provided an outstanding educational resource. The local-food movement is growing in rural Maine, and its relative availability is a great asset for the college and the community.

Because of our campus's location and educational mission, it made good sense for the college to become a thriving center for initiatives in local food growing. On one lovely spring day, in the early days of my presidency, we held a local-foods gathering in our student activities building, featuring farmers and providers from the local community. That evening we hosted a wild-game dinner at the college (an annual tradition). I attended both events, sandwiched around a meal in our cafeteria. Although our food service was independent and was owned by the college, it imported much of its food from national distributors. On that day, I had three campus meals, one in the cafeteria, another at the local-foods gathering, and a third at the wild-game dinner.

Michael Pollan's book *The Omnivore's Dilemma* had just come out, and I was deeply immersed in it. Pollan explores the origins of different food systems, simultaneously researching the natural history and the political economy of common meals. He describes industrial farming, "pastoral" organic agriculture, and "personal" hunting and gathering. It occurred to me that in the course of my three meals that day I had experienced all of Pollan's prototypes. The cafeteria meal consisted mainly of products of industrial farming. The local-food meal was made up of products of organic agriculture. The wild-game dinner consisted almost entirely of foods from the forest. I thought long and hard about the implications of these three meals and their meaning for Unity College. How could we make community food production and education a key feature of our sustainability efforts? How could we minimize our reliance on industrial farming? How could we build a deeper awareness of food into all aspects of campus life?

Unity College has a superb bioregional asset: a community that wholeheartedly supports sustainable agriculture. This was the ideal bridge between the campus and the community. Why couldn't the college become a leader in teaching, learning, and implementing sustainable agriculture? This could support interesting transformations in campus infrastructure (landscape design), curriculum (a new major in sustainable agriculture), and community partnerships (working with local growers and regional food banks), as well as college philanthropy (many interested regional donors). This was also an appropriate cultural fit, reflecting the spirit and the history of both the campus and the community. Emphasizing sustainable agriculture at Unity College could become a foundation for our sustainability initiatives.

Campuses throughout North America are taking the concept of sustainable food seriously. Hundreds of institutions are growing more food on campus, working with their food services to ensure the provision of healthier and more local food, and emphasizing the curricular possibilities of integrating sustainability initiatives with an understanding of food systems. Whether a campus is urban or rural, in a cold climate or a warm one, resource strapped or well endowed, there are dozens of opportunities for campus food-growing initiatives, while involving students, staff and faculty in the process. Similar to the growth of renewable-energy installations on campuses, in the last decade we have seen an extraordinary and inspiring increase in the availability of local food. Unity College is more than a participant in a regional food network. It is a contributor to a growing nationwide network of campus food awareness.

In this chapter, I'll cover the implications and potential of food for sustainability initiatives. Much as I did with energy in chapter 1, I'll begin by considering food from a naturalist's perspective. We can enhance our appreciation of food by observing its significance in just about every ecological relationship. An understanding of food is also crucial for enhancing understanding of indigenous culture and local natural history. I'll discuss the importance of eating well and why the food served on a campus sends a powerful curricular message to the entire community. That leads to a discussion of the relationship between food and campus culture. I'll urge that campuses undertake a food action plan that corresponds to and supports the climate action plan. In so doing, I'll suggest ways that campus leadership can lead the way in these efforts, while explaining how a campus food landscape can dramatically improve the campus aesthetic. I'll conclude by describing how sustainable agriculture has the potential to integrate the campus and the community in ways that improve the quality of life for an entire region.

Perceiving Food as a Naturalist Would

The best lesson I ever received in natural history came from reading Charles Darwin's October 3, 1835 entry in *The Voyage of the Beagle*. Darwin writes in great detail about his observations of Galapagos tortoises, noting their habitats, behaviors, and physiology. Darwin's acute observations were derived from a rigorous investigative protocol, connecting all aspects of species, niche, and habitat, anticipating an ecological methodology that wouldn't be fully elaborated for another century.[2] What do these tortoises eat? Where to they go to find food? Is there anything that eats them? Darwin writes:

I opened the stomachs of several, and found them full of vegetable fibres, and leaves of different trees, especially of a species of acacia. In the upper region they live chiefly on the acid and the astringent berries of the guayavita, under which trees I have seen these lizards and huge tortoises feeding together. To obtain the acacia-leaves, they crawl up the low stunted trees; and it is not uncommon to see one or a pair quietly browsing, whilst seated on a branch several feet above the ground.

As for what eats them, Darwin notes that "the meat of these animals when cooked is white, and by those whose stomachs rise above all prejudices, it is relished as very good food."[3]

Observing the food chain is an obvious way to understand ecological relationships. Yet this basic observational truth never really hit home for me until I read the aforementioned passage. With this insight I became a much better birdwatcher. If you want to find birds, figure out what they eat and go where their food is most likely to be. Insects and seed blooms are a good start. On every conceivable level, from observing soil micro-organisms (under a microscope) to tracking charismatic megafauna, or closely watching the feeding behaviors of seagulls, mice, and pigeons, you'll get your best natural history lesson when you understand the intricate food webs that all species weave.

Pollan considers the food chain in *The Omnivore's Dilemma*:

Ecology also teaches us that all life on earth can be viewed as a competition among species for the solar energy captured by green plants and stored in the form of complex carbon molecules. A food chain is a system for passing those calories on to species that lack the plant's unique ability to synthesize them from sunlight.[4]

Investigate all the rhetoric surrounding sustainable agriculture, slow food, and nutritious diets, and you arrive at the raw, organic, biological bedrock layer of organismic truth, now a trite bumper sticker, but true nevertheless: You are what you eat. The writer Gary Nabhan is a great exemplar of this. In a series of books covering subjects as diverse (but connected) as pollination, biodiversity, foraging, indigenous agriculture, seed dispersal and domestication, hunger and famine, and culinary excellence, he finds the essence of a bioregion in the place where natural history, food and culture meet:

The taste of the wild oregano somehow echoes the taste of the desert itself, for although its soils may be poor in water and certain nutrients, they are rich in the things that build flavor and, perhaps, character. When we eat crushed leaves of oregano in a vinegar-and-oil salad dressing or a hard-crusted dinner roll, we are indeed tasting the desert's essence—the challenges of being green (or gray) in a water-starved land, and the miracle of making it work on the land's own terms.[5]

I can't think of a better opportunity for studying sustainability and natural history than taking a foraging field trip. If you want to understand food better, go to the nearest park and find some.

This is not a romantic Euell Gibbons testimony to the virtues of stalking wild food.[6] It is merely a call to sustainability common sense, ultimately linked through curriculum to campus food and wellness. How we eat tells us a great deal about humans and the ecosystem, both personally and collectively, and some of our best lessons in sustainability, ecology, and human health are derived from understanding the natural history of food. We can experiment with this approach in our daily lives by paying more attention to the intricate food webs in any ecosystem. Whether we are in a New York City skyscraper, in a high-rise residence hall, or walking in a park, it is instructive to observe where organisms get their food. It is equally helpful to apply this approach to our daily meals. Can we trace the natural history of this morning's breakfast? This is a primary curricular challenge for campus food-sustainability initiatives.

Here is where things get more complicated. And this is where Michael Pollan's work gets particularly important. On most campuses, and indeed in all aspects of our lives, most of our food comes to us via industrial farming. Why is this significant for how we understand sustainability? It is helpful to consider Pollan's assessment of how industrial farming changes our perception of food. First, it reorganizes the food chain so that it revolves around fossil fuels, the energy sources that fuel intensive food processing and distribution. The energy budget of industrial agriculture is ultimately determined by oil. By understanding this, you grasp how there is a correlation between food awareness, energy systems, and climate action planning. Second, the processes of "raising millions of food animals in close confinement . . . and feeding those animals foods they never evolved to eat" in combination with the copious amount of high-fructose corn syrup in our foods means that "we are taking risks with our health and the health of the natural world that are unprecedented." Third, Pollan suggests, "the way we eat represents our most profound engagement with the natural world." Eating at McDonald's is a cultural indicator of how we engage in human-ecosystem encounters. Fourth, and what Pollan suggests is "most troubling, and sad, about industrial eating," is "how thoroughly it obscures all these relationships and connections."[7]

Pollan elucidates these relationships by building his narratives around the ecological origins of the industrial agriculture food chain. His book is an outstanding curricular device for understanding the relationship between food and sustainability. Ultimately, the natural history of food is where Pollan begins and ends. And so should we. As he suggests, "how and what we eat determines to a great extent the use we make of the world—and what is to become of it." He reminds us that it might seem like a burden to bring this rigorous approach to our daily eating habits, but the rewards are satisfying. There are ecological, political, and ethical reasons to eat with more awareness. From an educational perspective, food awareness is where human health most directly encounters the ecosystem, and it is an accessible way to link natural history to human flourishing.

When a college grows more food on campus, engages the community in the growing, serves local food in its cafeterias, provides comprehensive nutritional information, and connects all those initiatives to energy use, natural history, and human health, it can dramatically improve the quality of life for the entire campus community.

Eating Well

It isn't easy to eat well. It can be inconvenient and time consuming. Busy people tend to eat quickly. There are good reasons why fast-food restaurants and convenience stores are as successful as they are. Even Whole Foods markets and similar establishments cater to customers who want to purchase prepared meals quickly and conveniently.

The slow-food movement is organized around decompressing the daily meal. Prepare your food more slowly and deliberately. Take care in choosing the ingredients. Enjoy your meals. Make them convivial. Chew slowly. Great advice, we think, but when will we have time to eat this way? We rush our students (and ourselves) from classroom to meeting to activity and then back again, doing our best to squeeze in a meal, and reserving "slow food" for relaxed, recreational dining experiences, when we can "afford" to do so. It is also challenging to provide local, whole, and organic food at competitive prices. In the short term, it is seemingly cheaper for campuses to contract with food services that are deeply enmeshed in the industrial agriculture system. And students (or anyone on a budget) might perceive more nutritious food as more expensive food, or as less filling. Let's face it: When we want comfort food, would we rather have macaroni and cheese, or steamed kale? A donut, or a pomegranate?

It is encouraging to observe the number of campuses for whom the provision of wholesome food is a priority. Many of the well-known corporate food-service providers are catering to these requests. I have had meals in quite a few campus eateries that featured an outstanding array of whole food choices. Some campuses have special corners within their cafeterias for this purpose. Some universities serve only food that is vetted through sustainability criteria. The University of California at Davis and the College of the Atlantic have accomplished this on two very different campuses. Unfortunately, many campuses are still offering industrial food meals. Nevertheless, we have come a long way from the food fights in *Animal House* to the organic salad bars on many campuses. In the next few sections, I'll outline some coordinated approaches for moving toward a more sustainable campus food system.

First a few words about eating well. Mounting evidence suggests that poor food choices contribute to obesity and diabetes, to a range of chronic health disorders, and to attention deficit, fatigue, stress, and anxiety.[8] It isn't immediately evident how the daily consumption of soda, french fries, and sugary breakfast cereals leads to these conditions. However, there is increasing interest in assessing the direct links between diet, psychological health, and academic performance. I will not go into detail on these issues other than to suggest that every campus should emphasize the practice of good nutrition as essential to its academic mission. There is no more important life skill for students than to understand how to eat well. Such awareness may improve student's and employees' health and performance, and may contribute to reducing the costs of health care and improving morale. Healthy eating may improve a variety of academic measures.

Nutritional science is making great strides in identifying the foundations of a healthy diet. There are many useful nutritional guides for eating well, any of which are immensely helpful for campus food strategies.[9] Although it is sometimes difficult to separate nutritional trends from nutritional facts, there are common-sense guidelines: Increase the consumption of fruits, vegetables, and whole grains. Reduce consumption of sugar, salt, and fat. I'm sure that as nutritional science advances we will discover new ways to combine nutrients, and even develop specific diets linked to multiple physiological and ecological variables.[10]

What matters most is that these approaches provide an ecologically oriented and sustainable approach to nutrition and health. They can easily be incorporated into both campus food systems and sustainability curriculum. Eating well should be a priority for all food sustainability initiatives.

Food and Campus Culture

I have had opportunities to eat meals in dozens of college and university settings. I carefully check out the kind of food being served, what students are choosing to eat, who they eat with, and how much food waste they are generating. I find that such observations go a long way in helping me assess the depth and effectiveness of food sustainability initiatives on campus.[11] The food services of the more progressive, environmentally oriented colleges, especially those with active sustainability programs, are more likely to feature local and organic foods. When sustainability initiatives are marginal, you are more likely to encounter industrial agriculture food regimes.

At Unity College, I found that, as the president, I could have a significant influence on all these matters. And I didn't have to wait for faculty votes or lengthy approval processes. It was more a matter of setting a good example, working with senior administrators, and clearly explaining why establishing a wholesome food regime should be connected to our sustainability mission. Even if we marginally increased the costs of food delivery, I thought, the short and long-term paybacks would improve the college. I also assumed that building an enduring food sustainability program required the cooperation of students, employees, faculty members, and board members. I took a two-pronged approach: initiate change where it is easiest to do so, then work the politics so the change endures.

Fortunately, as I have mentioned, Unity still had a college-owned food service. We resisted all attempts at franchising. Of course many food services have been increasingly accommodating to sustainability initiatives, but a college that has its own food service can have more control of the results. I made it clear to the Dean of Student Life and the Director of the Food Service that the cafeteria should serve a higher percentage of local and organic foods, to work with community farmers, and to grow more food on campus. I also suggested that the college use professional development funds to support the cafeteria staff, sending people to relevant conferences and training programs. I encouraged the Vice President for Advancement to cultivate donors who would support these efforts. We received several grants to support food-growing initiatives on the campus. We coordinated these efforts with local growers and with the regional food bank.

For every catered event, we partnered with local providers, letting them know that we wanted our food locally sourced to the extent possible. One of our caterers changed her entire approach to cooking, both for her business and her family. In the cafeteria and in the café, we provided visible and diverse nutritional options. We established ethnic food nights.

I made it a point to eat in the cafeteria regularly, taking great interest in the food that was being served, speaking with the employees about why these efforts were important, and complimenting them on their successes.

Though we emphasized food awareness among students and staff, we never went overboard with these changes. We didn't get rid of the french fries and pizzas. We still used some food distributors that relied on the industrial farming system. Before too long, the campus food culture changed and even more new initiatives emerged from the grassroots. The role of campus leadership is to provide a venue so that these changes are supported, encouraged, and rewarded.

Whenever I visit a college campus, the sustainability advocates wish to discuss their campus food system. If you were to compare college and university campuses of ten or even twenty years ago with what is available today, you would notice a great change in the availability of wholesome and local food. We still have a long way to go in integrating all aspects of campus food delivery—from the growing to the consumption. And a great deal of research remains to be done—research on the relationships among nutrition, academic performance, and stress, on food quality and quality of working life, on food awareness and sustainability life choices, on how food choices affect climate action planning, and on whether the provision of wholesome campus food affects the food behaviors of students, staffers, and faculty members. In the meantime, we should continue to support low-cost, sustainable approaches for transforming campus food systems.

Food Action Planning

If a campus intends to transform its food system, it should consider engaging in a comprehensive, participatory planning process. To maximize inclusion and support, such a process might involve multiple constituencies—students, senior administrators, board members, and community partners. Growing more food on campus, supporting sustainable agriculture, changing food habits and behaviors, and building nutritional awareness into the fabric of student life and into the curriculum are significant change processes requiring coordination, coherence, and timing. A comprehensive food action plan can transform the health, vitality, landscape, and visual appeal of a campus. But it requires strategy, leadership, and behavior change. On many campuses, this process is generated in partnership with the student body.[12]

For example, Real Food Challenge is a student-based initiative that encourages campuses to develop a multi-year action plan, signed by the president, coordinating a full-fledged sustainable food effort. You begin with a campus food systems working group that includes "input from multiple stakeholders such as dining staff, workers, sustainability offices, NGO partners, relevant student groups, and student government."[13] Then you initiate an operations overview, assess current food procurement data, review best practices and achievements, and consider challenges or obstacles. This leads to a multi-year action plan with objectives, metrics, and timelines, linked to food procurement and supply chains, operations and facilities, contract processes, community involvement, and various campus food initiatives. Such an action plan requires a strategic policy initiative, derived from a campus-wide discussion of food sustainability, and finally connected to the mission and values of the institution. It necessitates support from all of the senior leadership, as it affects finances, student life, and community partnerships.

The website of Real Food Challenge has a roster of participating colleges and universities, with a summary overview of their efforts, aspirations, and challenges. Nearly 400 institutions are participating in this program, coordinated by student volunteers. As of this writing, the protocols and procedures for food action planning are not nearly as comprehensive as the American College & University Presidents' Climate Commitment. However, there is every reason for a university to devote as much attention to sustainable food initiatives as it does to renewable energy. The effect on campus life can be equally significant.

Just as executive leadership can provide a huge boost to sustainable energy efforts, it can stimulate and encourage food action planning. In fact, taking tangible steps toward food sustainability is much less complicated financially and is likely to be less controversial. It is much cheaper and easier to grow food on a campus than it is to finance major energy retrofits. Moreover, it is healthy and virtuous to serve high-quality food. A campus of any size, in any region, can rally multiple constituencies (from all political and educational perspectives) around these efforts.

Sustainable food and energy processes are synergistic. An increased emphasis on local food can potentially reduce an institution's carbon footprint, lessen its ecological impact, and minimize waste. Such initiatives, in partnership with the local community, can provide a boost to "green" businesses, while saving the campus money, promoting regional self-reliance, and allowing more control over costs and expenditures. Finally, food and energy initiatives contribute to an ecologically oriented campus environment.

There are good examples of campus food services cooperating with food sustainability efforts.[14] When campuses use corporate food services, they can negotiate their contracts to emphasize their locally derived food sustainability philosophy.[15] Good examples of food sustainability efforts can be found at the University of California at Davis, at Cornell University, at the University of Montana, at the University of Maine in Orono, and at the University of Minnesota. Many of these initiatives work in parallel or cooperation with more traditional agricultural research projects.

The Campus Food Landscape

Most campuses that emphasize sustainability have active, community-oriented farms or gardens. It is inspiring to observe the number of campus food-growing initiatives that are sprouting around North America. From inner-city campuses to remote rural outposts, from rooftop gardens to extensive organic vegetable plots, from permaculture to greenhouses, you will find dozens of institutions that are engaging in such programs. As recently as the year 2000, you would have been hard pressed to find such activities except on campuses with agriculture programs. Food-growing initiatives are now mainstream, and they are happening in every region of North America. Many campuses proudly show off their farms and gardens. Some are adjacent to the campus; some are centrally located.

Growing food is a relatively inexpensive way to beautify the campus grounds. At Unity College, we placed gardens wherever we could, essentially in any area that had sufficient space and light. We grew vegetables for the cafeteria and the local food pantry. An enterprising student worked with a faculty member to construct half a dozen wildflower gardens that bloomed throughout the growing season. These gardens positively upgraded the campus aesthetic. They also enabled us to dramatically reduce campus mowing. Extensive food growing at Unity College completely changed the campus landscape, and it had a profound effect on the campus culture. Active gardens convey a sense of belonging, a deeper relationship to place, and an impression of involvement. A garden-filled campus presents an image of hands-on learning, cultivates an ethic of care, and reflects a sense of hard work and diligence.[16] This is the tangible, on-the-ground, real-time effect of food action planning.

Food sustainability can have a significant effect on the quality of life on a campus. At Unity College, I derived great satisfaction watching students, staffers, faculty members, and community members working together in the gardens. We worked with the Maine Organic Farmers and Growers Association, with Veggies for All (a local community garden project), with the local food pantry, and with several foundations to build campus and community gardening infrastructure. These projects not only contributed to our cafeteria menu; they also provided support for local elementary schools, regional hunger alleviation, and combined efforts with local volunteer organizations (Unity Barn Raisers and Unity Rotary). Anyone could take an hour or an afternoon to work in the garden, chat with community members, relieve stress, and engage in community service. Just about every sustainably oriented campus has similar examples, tailored to the specific region and culture of the college. Kingsborough Community College in Brooklyn has an extensive urban gardening program. Furman University in South Carolina has a lush vegetable garden prominently placed next to the Shi Center for Sustainability. The University of Arizona has gardens designed for water-stressed urban environments. Pick any region of the country and you will find examples of interesting projects that involve the students, the community, and a range of volunteers and researchers.

Growing food improves the campus ecological environment. When campuses prominently grow food, or cultivate edible landscaping, they are engaged in more than just gardening; they are making an ecological statement about integrating humans with the ecosystem. Similar to energy initiatives, or any of the other projects described in this book, campus food growing makes a public statement about the meaning and purpose of education. Visitors to the campus share ideas about sustainable gardening practices. The campus becomes a center for the practice of regional sustainable agriculture and a public demonstration for sustainability actions and practices.

From a landscape perspective, why limit ourselves to food growing? The campus can establish butterfly gardens, bird migration stops, habitats for wildlife, and experimental plots for endangered flora. Just as the process of growing food can explore heirloom seeds and regional varieties, the entire campus can emphasize biodiversity, pollination, and other ecological concepts. Gardens and food growing are a great place to start, and in effect, they contribute to campus biodiversity. With visionary leadership and coordinated action, the campus food landscape can become a prototype for multiple approaches to ecological experimentation. The hopeful news is that these projects are already happening.[17] It is incumbent on sustainability advocates and practitioners to explain why they are important, and how the cumulative effect of so many of these projects nationwide can have a profound influence on how people perceive their place in the biosphere.

I'd like to see a nationwide campus data base that visually connects the new food landscape. This would include photographs, maps, planting tables, seed atlases, and other relevant data. Perhaps these projects taken together will become campus food corridors that simultaneously support local agriculture, biodiversity, community service, and nutritional awareness. These can be coordinated with curricular approaches, linking colleges and universities to local farmers, health-care facilities, and K–12 schools. The cross-sector ramifications are exceptional. The groundwork for this integrated food landscape has already been laid. Now it is a matter of creating a seamless network of these relationships.

Food as Curriculum

Peter Menzel and Faith D'Aluisio, in their illustrated book *Hungry Planet*, portray the ecology, culture, and anthropology of global food systems. They do so by combining a series of contributor essays with specific portraits of diverse families and cultures, thus linking family meals to the political ecology of food. In a succinct two-page summary of global food history, the environmental historian Alfred Crosby suggests that "the number of us who suffer from the diseases of overeating may be, for the first time in history, approaching that of those suffering from undereating. It may be that we will suffer the former problems to the extent that we cure the latter. Food was, is, and will always be central to the story of *Homo sapiens*."[18] This observation and many others in *Hungry Planet* generate poignant curricular challenges. What *Hungry Planet* accomplishes for its readers, a campus food curriculum can model for its students. Consider just a short list of relationships that can be explored by deepening food awareness—tracing the natural history and political economy of a daily meal, analyzing the nutritional contents of a meal, charting personal eating habits over the course of a week, assessing the social experience of eating, observing how much food is wasted during an average meal, determining how much energy it takes to cook a meal, developing an organizational flow chart depicting food procurement decisions, connecting diet and nutrition to psychological moods and well-being, comparing the availability of food through geographical space and historical time, or interpreting the relationship between food and opportunity.

Innovative faculties develop syllabi and research programs emphasizing these questions. Typically such teaching and research is organized around formal classroom learning. Yet each of these relationships reflects the daily life of the campus food system in a way that goes well beyond the boundaries of campus.[19] With just a little bit of imagination, good discussions around these topics can be embedded in campus cafeterias and eateries. The campus food service is underutilized as an educational resource. We tend to stereotype food services as administrative providers. We neglect their powerful educational role. I highly recommend that campuses develop sustainable food educational outreach committees, with representation from the food service, the department of student life, the faculty, student groups, and the sustainability staff. A primary task is the development of interpretive signage specifically designed to address some of the topics listed above. Such a program can be combined with formal coursework, student life activities, and community outreach. What a superb way to engage community partners, including local food providers, restaurants, distributors, public health officials, physicians and nutritionists, and people from diverse cultural communities. Does the campus provide suitable whole foods cuisines and explain why it does so? Is there adequate information provided about the food—where it comes from, its nutritional value, and why it is being served? Does the campus build food awareness, including nutritional concepts, into its curriculum? Perhaps most important, does the cafeteria provide an interesting learning experience? Any sustainable food initiative is incomplete unless it incorporates an educational agenda.

The Food Network

On a visit to the University of Kentucky I had the good fortune to attend a community breakfast sponsored by the Sustainable Agriculture
and Food Systems working group. First Friday is a networking forum
that brings community members together with students, staffers, and faculty members. As I entered the gathering place, I took my spot on a long
line of people waiting patiently for their "local foods" breakfast of om-
elets, vegetables, and bread. I picked up my breakfast (served by a faculty
member) and then found a seat at a long family style table that seemed
to stretch the length of the room. Well over a hundred people attended
this breakfast. They were enjoying a remarkably festive morning. After
breakfast, there were several short presentations by local folks, including
a couple that abandoned their city life and started (without any experience) a cheese-making farm.

This breakfast gathering was a wonderful testimony to the power of
building an inclusive community food network. It was clear that the research efforts of the UK Sustainable Agriculture and Food Systems Working Group were deeply intertwined with community agriculture. The
campus was a hub with spokes radiating in every conceivable direction,
culminating in a breakfast meal and good conversation. Gatherings like
this, many of them organized on college campuses, reflect a grassroots
network of sustainable agriculture. Many campuses are using food networks such as these to build resilient community relationships. It is a very
effective way of building "town-gown" relationships, using the campus as
a community resource, and engaging students in service learning opportunities. At Unity College we had numerous community dinners, organized
in partnership with regional food, hunger, and sustainable agriculture organizations. We developed a constructive relationship between these organizations and the campus. Nationwide and globally, there are hundreds
of examples of such efforts.

To get a sense of how extensive and embedded such networks can
become, and to illustrate why college campuses are ideal network hubs,
consider the long list of organizations that are involved in the farm-to-
college programs and other food system activities of the University of
California at Santa Cruz: the Food Systems Working Group, the Center
for Agroecology & Sustainable Food Systems, the Monterey Bay Organic

Farmers Consortium, the Agriculture Land Based Training Association, the Community Alliance with Family Farmers, the Program In Community & Agroecology, the Community Agroecology Network, Students for Organic Solutions, the Education for Sustainable Living Program, and the Santa Cruz County Food Systems Network.

Although campuses are excellent meeting grounds for these networks, some programs also emerge independent of higher education. Many cities have prominent sustainable food system networks, featuring farmers' markets, farm-to-cafeteria school programs, partnerships with restaurants, adult education venues, and food justice literature. Among many others, check out the Detroit Black Community Food Security Network, the New York City Food Systems Network, and the New Orleans Food and Farm Network.[20] In these settings and many others, campuses can work with already existing networks, providing additional support, especially in terms of volunteer help and research prospects.

The campus can exercise leadership by exploring how sustainable food networks can build cross sector partnerships. It makes great sense to work with health-care facilities, social welfare agencies, K–12 schools, and even museums, parks, and recreational facilities. Similarly, there is no reason why campus approaches to food sustainability should be located mainly in sustainable agriculture programs. Why not engage a full gamut of potential partners, including colleges of education, medicine, planning, architecture, and the humanities? Food awareness is an effective, and relatively non-controversial way to bring sustainability partnerships into the daily academic life of an entire campus, while simultaneously supporting diverse community organizations.

3
Materials

The Greenest Cleaning Materials

On my strolls around the Unity College campus, I sometimes struck up conversations with people I didn't ordinarily encounter in meetings. I made it a point to check in with employees of the public safety and facilities departments and the cafeteria. The conversations would always start innocently, usually with our observations regarding the weather, or the Red Sox, or what was happening on campus. Although employees treated me differently because of my position, I did what I could to break through such barriers. Over time, people opened up, and would offer their opinions on just about anything related to Unity College.

Everyone knew that I was interested in campus sustainability. However, I also knew that it would take time before the staff understood what that meant for Unity College and for their jobs in particular. I couldn't gauge, especially in the early days, whether this was just my initiative, supported by a few advocates, or whether the idea was really taking hold.

One cold and cloudy Saturday in November, when very few people were outdoors, I encountered a man named Keith who had recently been hired as a custodian. I had met him before, but we hadn't spoken much. I asked him how he was doing and whether he was enjoying his work at the college. After a few moments, he told me that he had a dream for the college. I asked him to tell me about it. He said that he wanted Unity College to use "the greenest cleaning materials" of any campus in the state of Maine. I was moved by his comments. I thought the sustainability message might really be hitting home. I told Keith that I would do everything I could to support his dream.

Then I wondered: Exactly what are "green" cleaning materials? Do Keith and I mean the same thing? What measures and criteria are we using? Does Keith's supervisor endorse his view? How far can Keith take his dream? Would the entire college go along with it? How might I be most helpful? How do we expand the idea of "green materials" to include all construction projects, procurement decisions, waste removal and recycling policies? How can we link whatever policies we develop to curricular objectives? Who will take this on?

This was just one example of how the college had to assess its use of materials. One of the chemistry professors was desperately trying to clean up some archaic storage closets. We were trying to figure out how to buy "green" chairs and desks, how to make sure we used toxin-free paints and carpets during remodeling jobs, and how to get rid of mold.

I asked the Sustainability Director to speak with Keith and his supervisor, to include the Vice President of Finance in the discussions, and to sketch out a policy on materials procurement for the campus. I asked the Human Resources Director to work with the chemistry professor to set strict and reliable safety standards for the lab. I suggested that all these policy recommendations be vetted by our senior team so these projects would take their place as a whole-campus initiatives.

As the president of a small, resource-strapped college, I wondered how far I could take any initiative before I would drive people to the limits of their work capacity. It took time to approach these challenges systematically, and I never felt that we had the most rigorous criteria for making our decisions. Still, we persevered, building a spirit of compliance, and a much greener workplace as a result.

A university's use of materials reveals an array of ecological, cultural, economic, and political relationships. In this chapter I'll suggest a sequence for conceiving how a campus can move toward more sustainable approaches to materials. I'll begin by looking at materials from a naturalist's perspective. How do we reinterpret the use of materials so that it inspires a rich understanding of the natural world? Then I'll look more closely at the lineage of sustainability, how it is derived from a critique of consumption, how it incorporates concerns about toxicity, and why it emphasizes the "reduce, reuse, and recycle" approach. I'll discuss how balancing concepts of scarcity and abundance broaden our perspective on the use of materials. I'll review some impressive new tools for better understanding the ecological impact of the use of materials, specifically the idea of an ecological "footprint." How practical are these approaches, and how might they be implemented throughout a campus community?

A sustainable approach to materials science also suggests that we add three more R's to the equation: restore, redesign, and regenerate. Some universities are researching new approaches to the use of materials that emphasize durability, resilience, modularity, and adaptability. We have as much to gain from the science of "green" materials as we do from traditional concepts of conservation. I'll conclude the chapter with a brief survey of the new field of sustainable design, a convergence of ideas regarding chemistry, craft, ergonomics, human behavior, marketing, and ecological aesthetics.

The Magic Marker

I have an enduring childhood memory of my first encounter with a Magic Marker. In a second-grade classroom of the 1950s, children paid close attention to art supplies. In those days a small box of Crayola crayons was standard. The epitome of affluence was having access to one of the bigger boxes—one with 48 or 64 colors. Exploring the colors was as much fun as reading their names. Who from that era can forget the mysteriously named "burnt sienna?" One day Miss Wilbur used a new writing implement to embellish her lesson. She called it a Magic Marker. None of us had ever seen one before. In the beginning, they were black, and they were hard to come by. Within a few years they were commonplace. Now there are many millions of marker pens in schools, households and workplaces throughout the world.

In my early years of teaching environmental studies, I had a favorite introductory activity. I would ask the students to break up into small groups; I would then give each group a handful of marker pens and some paper. Their task was to tell me everything they could about the origins of the Magic Marker and to represent their knowledge as a diagram, drawn with the supplied markers. I assigned this activity hundreds of times in a variety of curricular formats. I asked them to trace the ecological/economic pathway of the markers. What were they made from? Where were they made? How were they transported from source to user? What happened to them when they no longer were useable, when the magic was gone? In all the years of using this activity, I never found anyone who could describe the process with any level of specificity. Rather, I received clever diagrams that resembled creation myths—the markers were made out of plastic, derived from petrochemicals, originating in the Carboniferous, and then assembled in a factory somewhere, and delivered through a supply chain.

This was an interesting curricular approach because it evoked questions about the production, distribution, and consumption of an ordinary consumer item, starting with a symbolic name—Magic Marker—that conveyed the mystery, or at least the lack of consumer knowledge regarding the product. So the activity sparked some good conversations about the complexity of the supply chain, the ecological and human effects of the production process, the dynamics of extracting natural resources, the global distribution process, the proliferation of the markers, how they were differentiated by marketing, and then assessments of their ultimate and relative value. We would discuss how much it was necessary to know, not only about the Magic Marker, but about any consumer item that we use. Should a responsible "environmental" citizen be able to trace the ecological/economic pathway of any commodity?

It isn't only the marker that's magic. Almost every consumer product is the material embodiment of an extensive, global industrial production system, typically entailing complicated commodity exchanges, multiple production and fabrication sites, requiring energy, chemical processing, automated and human labor, and resulting in ecological impact. It is astonishingly difficult to trace the natural history basis of most everything that we use, sometimes even food, shelter, and housing. This is the conceptual challenge of better understanding our ecological footprint.

Where do consumer commodities originate? This question is now fundamental to the sustainability ethos. We understand that every individual (or campus) purchasing choice is way more complex than it initially seems. One of the great advantages, even pleasures, of consumer-oriented society, is the provision of a great variety of goods. We tend to be most concerned when the price of a good becomes prohibitive, and it is typically at that point when we ask questions about the supply chain. Yet the "secret life" of stuff tells a complex, often global story of ecosystems, people, working conditions, and the management of natural resources. The sustainability movement promotes a deeper awareness of the relationship between consumer choices and ecological impact. Our use of materials—the Magic Markers that are used for teaching purposes, the athletic uniforms provided to campus teams, the chemicals used to clean campus classrooms and dormitories, or the materials used in construction processes—is therefore deserving of close attention.

Consumption, Abundance, and Simplicity

Questioning the aforementioned relationships informs the three R's of sustainability—reduce, reuse, and recycle—and presents a bold challenge to the presumed abundance and disposability inherent to a consumer society. For several decades, advocates of sustainability have suggested that there is a dark side to excessive consumption. How do consumption practices affect ecosystems, generate toxicity, and disrupt human and natural communities? Much of the sustainability agenda, as influenced by the history of environmentalism, calls attention to environmental pollution—from the "dark, satanic mills" of the industrial revolution to Rachel Carson's book *Silent Spring* to recent accounts of industrial pollution in China. The logic is that if we reduce consumption and carefully manage natural-resource extraction, then we can also diminish these negative effects.

This leads to a generalized critique of the affluent society, dating back to the postwar era, that questions the presumptions of affluence, suggesting that materialism doesn't lead to happiness, that it jeopardizes environmental quality and community life, and that if we reduce our consumption we will have more time to enjoy what we have. There are, as William Leiss wrote several decades ago, "limits to satisfaction" beyond which we are trapped in an endless loop of a search for elusive happiness, mistakenly linked to identification with consumer goods.[1] By reducing our consumption, reevaluating our habits of consuming, and paying more attention to what we want and what we acquire, we can minimize our effect on the planet while maximizing our enjoyment of life.

Reducing, reusing, and recycling is an ideal policy and behavior response to this challenge. Colleges and universities are taking the lead. Recycling is a relatively simple way to call public attention to reducing the use of materials. We don't have to worry so much about the ecological impact of the things we use if we use them over and over again. Reducing, reusing, and recycling, in combination, summon a variety of potential behaviors and results, including frugality, waste reduction, and conservation. Recycling awareness also raises questions about the life cycle of materials, prompting a deeper regard for durability, decomposition, and dependability. On a college campus, this is where sustainability advocates, facilities people, and budget makers often work together, and where they get support from students and from faculty members. The various national recycling competitions have caught on because they are fun and the participants feel as if they are accomplishing something. Constructing ecological footprints is a systematic curricular application of the three R's. A class or an entire campus can trace the intricate life cycle of materials, raising awareness of their ecological origins and costs.

However, sustainability criteria are often perceived as parsimonious. The three R's, though intended to promote mindfulness about consumption and the ecological footprint, also reflect a "limits to growth" attitude. Many environmental solutions (from population reduction to limiting resource extraction) are based on a post-Malthusian scarcity model. This model may seem unnecessarily restrictive, even austere and possibly dismissive of global poverty challenges.

The three R's prompt a fundamental philosophical question concerning the prospects for human progress. One way to consider progress is to establish criteria for human (and planetary) well-being and then determine whether things are getting better. The economist Charles Kenny, in a book titled *Getting Better*, outlines how global development contributes to an "historically unprecedented improvement in health and education, gender equality, security and human rights." Yet he also acknowledges that the "taxing of the global commons surely raises concerns about the sustainability of the current trajectory of global development." He further suggests that "doubling the incomes of the world's poorest 650 million people would take the same resources as adding a little under 1 percent to the income of the world's richest 650 million." Kenny offers a "Hippocratic" approach to global sustainability. "The threat should be confronted," he writes, "but not at the cost of depriving poor people alive today of their access to basic human needs."[2]

I raise these issues because I believe that sustainability should emphasize the prospects of abundance as much as it warns of the threats from scarcity. The history of human life on earth is always a balance between these two extremes. We can promote human welfare while reducing ecological debt. We can project growth trajectories that simultaneously reduce the worst effects of resource extraction while investing in creative technological innovations. Sustainability should entail an optimistic outlook, utilizing the fruits of abundance while striving to diminish scarcity and optimizing human welfare and the preservation of the biosphere. The word "abundance" connotes a rich supply of goods and services, an overflowing fullness, affluence and wealth, generosity and surplus. "Scarcity" connotes insufficiency, shortage, and infrequency.

Abundance is a fitting perceptual model for sustainability. We are living in the early days of some potentially very powerful emerging trends that maximize collaboration, social networking, technical innovation, global communication and connectivity, open-source research, and grassroots service and philanthropy. These are the fruits of a genuine cognitive prosperity that provides people with unprecedented access to the production and consumption of information. Clay Shirky describes this glut of information as a form of "cognitive surplus" and the various ways people can transform their lives and communities by redirecting how they use this surplus.[3] Michael Nielsen describes how "networked science" is transforming the nature of scientific inquiry, promoting collaboration, discovery, and innovation.[4] Peter Diamandus and Steven Kotler explain how "technophilanthropists" are using microfinance, investment, and ingenuity to disseminate appropriately scaled and widely disseminated technologies that provide clean water, more nutritious food, durable materials, and safe energy.[5]

Why is this discussion of scarcity and abundance relevant to how we conceive of materials? It is emblematic of the tensions generated by the sustainability ethos. How do we balance the necessity of recycling with the prospects of plenty? How do we call attention to the ecological impact of profligate growth while investing in future technologies? We are on the verge of a new generation of ecologically sound materials designed using synthetic biology, industrial ecology, and perhaps genomic engineering or nanotechnology. A forward-looking approach to the use of materials will use the ecology-based three R's as the foundation for three more R's: redesign, restore, and regenerate. These six R's provide a more comprehensive, futures-oriented, design-based approach to sustainable use of materials, acknowledging the necessity of innovation and calling attention not only to how and why we use materials but also to how we make them and how long they will last. Before we turn to the prospects for these new technologies and the concepts that fuel them, let us briefly examine ecological footprinting and the associated methodologies, as they ultimately serve as the ecological conscience for campus-based sustainable design.

Ecological Footprints

Since the mid 1990s, advocates of sustainability have developed a set of materials-related measurement tools. There are tools for individuals who wish to better understand how their life choices affect the biosphere. There are tools for communities in which people wish to work collaboratively to design more sustainable approaches to community life. And there are tools for college and university campuses that wish to measure their sustainability progress and minimize their effects—most notably the STARS (Sustainability Tracking and Rating System) process. All these tools are helpful in that they link individual, community, and campus actions and life choices to a complex chain of effects. They provide a variety of metrics and criteria to enhance comparative assessments. Most important, they aspire to translate the abstractions of money, the natural world, and human choices into tangible actions and results. On campuses, these tools allow administrators to assess a range of procurement decisions, consider strategic planning for sustainability initiatives, and better understand how the choice of materials used in construction can be linked to these broader considerations. They allow educators to help their students better understand some of the fundamental patterns of sustainability, ecology, and human decision-making processes, as applied to the material foundations of campus life.

There are dozens of "footprint" scales and questionnaires, including very useful online quizzes and worksheets. Mathis Wackernagel and William Rees' book *Our Ecological Footprint* was the first comprehensive guide to "reducing human impact on the earth."[6] Jim Merkel's book *Radical Simplicity* challenges readers to overhaul their lifestyles and reject consumption while comprehensively calculating ways of reducing their footprints. The "backstory" approach is equally helpful, prompting readers to trace the narrative of commodities—the equivalent of the Magic Marker exercise. And in *Stuff: The Secret Lives of Everyday Things*, John Ryan and Alan Thein Durning tell the story of the everyday consumption of coffee, newspapers, T shirts, shoes, computers, and cola.

The STARS process adopts for campus use the best of the techniques used in various "footprint" scales and questionnaires. Because we are still learning how to apply these tools effectively and instructively, STARS is continuously revising its criteria and metrics. STARS is particularly pertinent as a tool for assessing materials use in the cases of computers, cleaning products, and office paper; it also offers a vendor code of conduct and comprehensive details on reducing waste, on recycling, and on the management of hazardous waste.[7] STARS is a work in progress. The tools will continue to improve as we better understand their usefulness and relevance and as more people become well versed in their use or come to accept them as a necessity of campus sustainability policy. Also, we will develop more specificity regarding building materials, lab equipment, vehicles, athletic clothing, and just about any campus commodity.

What is interesting and hopeful is that the sustainability ethos (along with other global economy considerations) is encouraging the use of durable, "green" materials. When they are more readily adopted by business, industry, consumers, and campuses, the market dynamics will reduce the prices of such materials. I suspect that campuses will soon have a range of innovative sustainable procurement choices, based on a new generation of "green" materials. The STARS program (or some new iteration) will necessarily have to evaluate the ecological footprint of these new materials. This will present new challenges and opportunities for campus procurement strategies. Perhaps campus research programs will lead the way by inventing, testing, and then assessing "green" materials design and manufacturing.

"Green Chemistry"

The most difficult ecological footprint challenge is tracing the intricate pathway of a campus's use of materials. What chemicals are embodied in these materials, and how should their relative toxicity, durability, and resilience be assessed? Many campuses have initiated comprehensive recycling protocols, while minimizing their waste streams, and reassessing procurement policies. These are all good starting points. If campuses nationwide could make significant progress in all these areas (and there are hopeful signs that this is happening), they will collaboratively reduce their collective ecological footprint. The more difficult implementation challenge is to minimize materials' toxicity while maximizing their performance. Paul

Anastas, the founder of "green chemistry," puts the challenge this way: "It makes no sense to be assessing one chemical at a time when we know there are over 100,000 chemicals in commerce today and that there are over 4,000 chemicals being invented or discovered every day, and when according to at least one Nobel laureate in chemistry, there's potentially ten to the 63rd new chemicals yet to be invented of modest molecular weight."[8]

One of the great contributions of the environmental movement, especially in the early days, and as spearheaded by Rachel Carson and Barry Commoner, was to call attention to the dangerous toxicities generated by many industrial and manufacturing chemicals.[9] This remains an enormous global and ecological health concern. It is where human health meets sustainability. But how can such industrial processes be adequately traced? It is difficult enough for a campus to fulfill its sustainability responsibilities without also having to scrutinize the materials that pass through the campus.

Of course, there are many useful regulations already in place that are designed to promote workplace safety and for dealing with hazardous wastes. But how can a campus oversee all the domestic and consumer products it uses? There are emerging protocols that can be very helpful, including assessments of putatively "green" building materials, cleaning products, office supplies, and consumer products.[10] Yet we still lack a standard process for evaluating these materials, and we need a better understanding of their durability and performance in relationship to their relative toxicity.

A very hopeful development is the growing "green chemistry" movement, spearheaded by Paul Anastas and first implemented at the U.S. Environmental Protection Agency. According to the EPA, "Green chemistry, also known as sustainable chemistry, is the design of chemical products and processes that reduce or eliminate the use or generation of hazardous substances. Green chemistry applies across the life cycle of a chemical product, including its design, manufacture, and use."[11] In partnership with the American Chemical Society, the EPA lists fifteen domestic and eleven global academic "green chemistry" programs. Most of these programs cite the twelve principles of "green chemistry" as the basis of their curriculum and mission. These principles emphasize reducing hazardous waste, designing safer chemicals, pollution prevention, energy efficiency, and innocuous degradation.[12]

In view of the significant impact of chemicals on the environment, and the extraordinary prospects for using chemistry and engineering as a foundation for materials design, it is surprising that more campuses aren't developing innovative programs in "green chemistry." Anastas emphasizes the potential of this approach: "Green chemistry is redesigning the matter that's the basis of our society and our economy and the materials that are used in all of the products we use so that it is more sustainable and less harmful to humans and the environment. . . . For every one process or product that's being reinvented using green chemistry and green engineering, there may be a hundred or a thousand yet to be rethought under these terms."[13]

More than posing a challenge for chemists and engineers, Anastas proposes a fundamental reconsideration of how we think about matter and energy as applied to materials. "Green chemistry" represents the scientific frontier of innovative use of materials—designing durable, non-toxic materials. It inherits the profound legacy of *Silent Spring*, redressing the pernicious consequences of the naive, poorly tested, and rampant use of toxic chemicals. Here is a pertinent agenda for sustainability, one that can immeasurably benefit people, species, and ecosystems. It involves complicated trade-offs, difficult discussions about motivation and profit, and fundamental differences about regulation. Nevertheless, it is a solutions-based orientation to minimizing the negative health effects of industrial and manufacturing processes. "Green chemistry" can improve the well-being of humans and that of the ecosystem.

Three More R's

"Green chemistry" requires a broad conceptual context to be most effective. The development and dissemination of "green" materials involves design as well as chemistry. Who is using the materials? For what purpose are they being used? What is their ultimate function? Are there aesthetic considerations? These questions, considered collectively, are the foundations of sustainable design, an approach to materials planning that weighs function, durability, efficiency, ergonomics, and beauty—the interface of ecology, aesthetics, and human behavior.

Consider a new LEED-certified residence hall. An architecture firm that emphasizes sustainable design will optimally engage its future residents in a planning process that maximizes energy efficiency, uses recycled materials, provides lots of natural light and air circulation, promotes privacy and convivial community gathering places, and emphasizes ways of reducing the ecological footprint of the building and its residents. But it is much less likely to be as rigorous about design and about the use of materials. For example, the LEED certification process, as helpful and effective as it has been, provides only general guidelines about how large a percentage of recycled materials is needed to earn LEED points. Are there desks, beds, and chairs in the rooms? What should they be made from? Will they be replaced in five years, or can they be built to last for a century or even longer? Are they designed to be visually appealing? Are they ergonomically suitable? One can go through the same process for any new construction or retrofit, whether of an athletic facility, a science lab, or an office building. A similar process can be initiated for any procurement decision. A life-cycle analysis isn't complete unless it asks such rigorous questions. Unfortunately we don't yet have a standard set of protocols and assessments for undertaking such initiatives.

There are two mutually reinforcing challenges and opportunities for campuses. First, they can promote programs that invest in cutting-edge sustainable design, working with chemists, engineers, biologists, artists, psychologists, and architects to develop new products and processes while providing the venues for using them. Second, and simultaneously, they can work with entrepreneurs, with developers, and with experts in organizational behavior and human decision making to promote cost effective and rigorous protocols for assessing these innovations. This kind of innovative research can emerge from campus sustainability initiatives. As a conceptual starting point, there should be a suitable way to build on the conservation-oriented slogan "Reduce, reuse, recycle" so as to emphasize innovation and design. Let us now add three more R's: restore, regenerate, and redesign.

Among the few people who understand the contents of consumer and/ or industrial waste are those who deal with it for a living and those who are impoverished and rely on it for survival. One of the great benefits of a consumer society is that waste is mainly disposed of by specialists. But this also results in a lack of understanding of the logistics of waste streams. One way to become more aware of waste disposal is to ask good questions about dissolution and disintegration. What happens to "stuff" when it breaks down? An innovative approach is to consider whether waste streams can be more than just innocuous—whether they can they *restore* toxic environments.[14] Are there chemical synergies, especially microbial relationships, that can promote mitigation, absorption, and/or cleanup? Are there locations on campus where such relationships can be researched and observed?

Such restorative processes will be significantly enhanced in the coming decades. The synergistic convergence of synthetic biology and genomic engineering will provide unlimited potential for ecologically sound use of materials. Synthetic biology is "the science of selectively altering the genes of organisms to make them do things that they wouldn't do in their original, natural, untouched state."[15] Genomic engineering involves the use of genetic programming to reconstruct organismic processes. As George Church and Ed Regis suggest in their book *Regenesis*, "genomic engineering will become more common, less expensive, and more ambitious and radical in the future as we become more adept at reprogramming living organisms, as the cost of the lab machinery drops, while its efficiency rises, and as we are motivated to maximize the use of green technologies."[16] Their vision is both compelling and daunting. "We are already remaking ourselves and our world, retracing the steps of the original synthesis—redesigning, recoding, and reinventing nature itself in the process."[17]

The ethical ramifications of these processes will be controversial for decades to come, and a full treatment of these questions is way beyond the scope of this book. However, sustainability science will inevitably involve using synthetic biology to *regenerate* "green" materials. Colleges and universities should develop partnerships with businesses and communities to consider the potential risks and benefits of regenesis. A truly progressive approach to sustainability will incorporate these challenges via curricular design and infrastructure experiments.[18] The emerging field of industrial ecology, with its systems-based, life-cycle approach to materials and energy flow, is providing a template for research and application. Inevitably synthetic biology, genomic engineering, and nanotechnology will contribute to industrial ecology. Which campus programs will provide the academic and research leadership?

Restoration and regenesis require a broad, collaborative, and participatory intent. What is the purpose of this new science? How will it serve human communities and promote ecosystem health? How can innovative material use become intrinsic to campus master planning processes? Enter the relatively new field of sustainable design, an approach to decision making that considers all six R's—reduce, reuse, recycle, restore, regenerate, and redesign—simultaneously, with ergonomics, aesthetics, and human behavior front and center. *Redesign* suggests that we take what we already have and frame our plans so as to incorporate sustainability criteria.

Sustainable Design in Practice

A good way to understand the possibilities of sustainable design is to briefly profile a model campus program. The Center for Sustainable Design Studies at Pratt Institute in New York has a remarkable portfolio of projects that emphasize the design and the use of materials. The CSDS's projects have a strong curricular component and are developed and implemented by members of the faculty in partnership with local business and community groups. Its online resource center is organized so that one can "integrate sustainable design practices into [one's] own work." The CSDS features a "material of the month" that assesses a range of "green" materials, including fabrics, paints, paper, foamboard, and insulation, providing a comprehensive report on all aspects of the products' life cycle. Its case-study profiles of products depict all aspects of a product's ecological footprint, including a map of the supply chain. The Pratt Design Incubator for Sustainable Innovation has "supported the launch of 23 companies and consulted for 15 organizations." Most important, the CSDS is neatly integrated into Pratt Institute's campus. "The CSDS encourages the use of our college campus as a living lab for innovation and establishes partnerships that provide students with applied experiences in sustainable design."[19]

Dozens of new sustainability-oriented programs have emerged since 2005. However, only a few of them (including those at Philadelphia University and the Boston Architectural College) incorporate "sustainable design" in their names. The Rochester Institute of Technology's Golisano Institute for Sustainability combines programs in architecture, resource recovery, remanufacturing, and sustainable production. There are several programs in industrial ecology, one at Yale University's Center for Industrial Ecology and several in Europe.[20] However, there aren't any programs that integrate synthetic biology, architecture, and sustainable design. I expect that in the next ten years we will see a new generation of interdisciplinary design-oriented programs that will entail various combinations of engineering, synthetic biology, and the arts.

Sustainable Design and "Green" Materials

How does a campus formulate a resilient and flexible policy for the use of "green" materials? A comprehensive assessment is time consuming and complex, and with the exception of various ecological life-cycle protocols (see chapter 5 below), there is no standard methodology for conducting one. However, as more institutions address the "materials" challenge, such procedures and methods will be developed. Convening meetings and conferences around the use of green materials is an ideal way to build working relationships with administration (finance, facilities, and planning) and academics (chemists, engineers, and architects), among others.

A campus can initiate a sustainable-materials policy by assessing its inventory, reviewing its procurements, and forecasting its future needs for materials. As a case study, it would be useful to develop a comprehensive materials inventory for a new campus building. The six R's can serve as a conceptual foundation for asking good questions and implementing ecologically sound procurement approaches:

Reduce According to criteria such as natural-resource extraction, water use, energy use, carbon footprint, and waste minimization, which building materials and artifacts have the smallest ecological footprint?

Reuse What are the prospects for using materials from other construction sites on the campus or in the community?

Recycle Do the materials used in construction have multiple purposes? Can they serve another function after they have served their original purpose?

Restore Do the materials used have the potential to contribute to the environmental quality of the site?

Regenerate Are there innovative, experimental "green" materials that represent new approaches to manufacturing, construction, and artifact development?

Redesign Are there ways of designing the building that can reduce ecological footprint, maximize sustainability awareness, and optimize the living experience?

Literature on sustainable design is proliferating, and increasing numbers of "green building" and "green architecture" firms are using the language of durability, resilience, modularity, adaptability, ergonomics, and craft. These concepts provides a great starting point for thinking about "green" materials and campus sustainability.[21]

Durability refers to how long something will last, how it holds up to repeated use, and its relative rate of depreciation.[22] *Resilience* is "the capacity to bounce back after a disturbance or interruption of some sort." Resilient design considers that capacity in determining which materials to use in buildings.[23] *Modularity* describes the replaceability, functionality, and versatility of a construction project or product.[24] Modular designs can be replaced, repaired and recycled more easily. *Adaptability* is "the ability of a structure to accommodate varied and often unknown future uses and changes with minimum of cost and effort."[25] This refers to both ecological footprint and financial investment. *Ergonomics* calls attention to the usability of buildings and materials, placing attention on how "performance, productivity, comfort, health and well-being are enhanced."[26] Sustainable design utilizes craft and its makers to develop ecologically sound, energy-efficient and artistic approaches to materials design and use.[27] Innovative *craft* research combines traditional materials with new digital technologies, and finds sustainable uses for synthetic materials.

Campuses have made notable strides in their thinking about sustainable infrastructures for food and energy use. There are hundreds of climate action plans, energy-efficiency initiatives, renewable-energy installations, food-growing options, sustainable-agriculture initiatives, and curricular programs that support them. However, use of materials raises new challenges, especially in view of how difficult it is to utilize ecological cost accounting in making sustainability assessments. In the coming years this will be an important consideration for sustainability programs. The emerging field of sustainable design provides a fitting conceptual framework for the planning, the creativity, and the analysis that will be necessary.

4

Governance

The Sustainability Inbox

Unity College's trustees had the good sense to send me to the Harvard Seminar for New Presidents just a week before I started my position. Of the many great lectures, activities, and discussions, one stands out. Called the "inbox activity," it was our very first assignment. The activity simulated the long list of tasks, expectations, and appointments awaiting a president first thing in the morning. Many of the messages were labeled urgent or had notes attached emphasizing their importance. This was a challenging exercise in time management and priority setting. There was no way one could fit everything into the coming day or even the coming week. One had to make on-the-spot decisions as to what mattered most.

Sure enough, during my first few months at Unity College I not only had an overwhelming inbox; I also had a long line of supplicants telling me what I should do, offering advice, and in some cases warning me (usually politely) of what might happen if I didn't heed their advice. People did this with good intentions, and with what they considered to be the best interests of the college in mind. I listened carefully to everyone, took their concerns seriously, avoided making irreversible commitments, and in the process I learned a great deal about the staff and the faculty. I tried to understand what mattered most to people, and how their concerns meshed with the specific challenges facing the college.

Inevitably there were some issues that required immediate attention, and I had to make some decisions using the best knowledge at hand. Yet it took me quite some time, perhaps a year, before I really understood how to manage my inbox. With experience I realized that there was a direct relationship between my daily inbox and the long-term strategic priorities of the college. This was my ultimate time-management challenge—how to coordinate what I was doing in the moment with what mattered most, and to communicate those priorities with my senior leadership team, the board, and everyone who cared about the college. Maintaining this clarity is never easy, especially given the "crisis du jour" that would so often cross my desk. Indeed, one of my earliest leadership challenges was defusing the crisis mentality that seemed to surround many people at the college, and building trust and confidence in how we ordered our priorities so we could make judicious and thoughtful decisions.

The inbox activity has many unfolding layers. Most colleges and universities place great value on transparency, accountability, responsiveness, inclusion, equity and participation—the basic tenets of good governance. Everyone wants to have access to the inbox in order to influence the college's priorities. The president's inbox, or that of any administrator, is seemingly of universal interest. But you also have to distinguish your own priorities as well. What is important to you and how do you ensure it remains a personal and institutional priority? What if, for example, sustainability is your priority? How do you keep it at the top of your inbox list, but also make sure that it rises to the top of everyone's inbox? How do you respect the tenets of good governance in making that occur?

The most compelling educational attribute of the inbox activity is its scale versatility in relationship to setting priorities. The "new presidents" at the Harvard Seminar were venturing into positions at every conceivable size and type of institution. Yet we agreed that the exercise was salient and profound. In retrospect we now know that the most enduring learning came on the job when we had to figure out how to lend our leadership voice, philosophy, and values to the changing circumstances of our colleges and universities. What does the inbox look like three or four years later? How do you use it to mobilize action, delegate responsibility, and coordinate strategic vision?

I'm particularly interested in the experiences of those presidents for whom sustainability emerged as a priority. What is clear from my discussions with colleagues at the Harvard Seminar and with dozens of presidents and senior leaders from many other institutions is that sustainability initiatives are most likely to have top priority for the whole institution when those leaders embody the ideal. Their values must be at the core of their priority system. When the values of sustainability are deeply embedded in a person's character and beliefs, then those values become intrinsic to governance and leadership.

The central theme of this chapter is the relationship between good governance and a sustainability ethos. What is leadership's role in coordinating that relationship? First, I'll discuss the relationship between sustainability, governance, and leadership. Are there values inherent in the sustainability ethos informing how we think about leadership? What are the moral components of sustainability leadership? Do they imply transformational change? If so, how might that be accomplished? Second, I'll review some practical "good governance" approaches for implementing sustainability initiatives. These include aligning strategic mission with campus priorities, building a sustainability leadership team, developing networks and partnerships, rewarding innovation and accomplishment, and synchronizing words and deeds. Third, I'll discuss some of the psychological challenges that inevitably accompany transformational change. How do you balance urgency and patience, boldness and compassion, innovation and tradition, autonomy and authority?

Although many of the examples are drawn from my specific experiences as a university president, I have found that the challenges I face are similar to those faced by many of my peers. The challenges of different institutions will always vary, and the political dynamics that inform leadership are always contextual. Nevertheless, the underlying patterns are often similar. And many of the sustainability practitioners I meet (many of whom are not presidents) face similar challenges. They are mainly interested in how to effect meaningful change and how they can be most influential as middle managers. This chapter is relevant to their concerns as well. Whatever your position or relative authority in a system, implementing change requires adaptive flexibility, social and emotional intelligence, an understanding of time management and priorities, and a well-articulated set of core values. Managing our sustainability inboxes is a collaborative and complex challenge.

The Challenge of Good Governance

Most colleges and universities place a huge emphasis on the qualities of good governance cited above—transparency, accountability, responsiveness, inclusion, equity, and participation.[1] A quick scan of the announcements of open administrative leadership positions in any journal of higher education will typically turn up a listing in which some combination of these qualities are stated as prerequisites for consideration. Administrative leaders are expected to listen well, delegate effectively, and act wisely. Colleges and universities typically pride themselves on promoting participatory processes. Civic engagement represents a noble (and commonly cited) curricular goal. Diverse constituencies—donors, students, alumni, faculty members, community businesses, politicians—promote their interests, all presuming access to leadership. In the bigger systems, governments, foundations, and corporations may have vested interests as well. They all want a seat at the table.

The governance challenge is magnified by the prominent role of faculty members. It is commonly understood that the faculty designs, stewards, and implements the curriculum. Insofar as learning is the central function of colleges and universities, faculty members rightfully presume that their voice as educators and researchers places them at the center of important decision-making functions. Yet most faculty members perceive themselves as having far less educational authority than they would like. And many administrators view faculty members as having more authority than they should. Sometimes a reasonable balance is achieved. However, there are many mitigating factors that create tension, especially issues related to revenue, investment, authority, and ultimately power. Colleges and universities devote an enormous amount of time to working through institutional politics.

Recently these challenges have become more intense because the cost of higher education is receiving much attention. The public doesn't understand why a college education should cost so much money. This becomes an acute issue in challenging economic times. Hence higher education is being more closely scrutinized for its perceived investment value, raising new concerns about accountability and access. There is even more pressure on campus systems for public reporting, cost accounting, and the fulfillment (or assessment) of educational goals and objectives. Such discussions often reduce the so-called value of education to economic criteria. Faculty members are now challenged to be more relevant. Administrators are now challenged to cut costs, increase revenue, and trim excess. This potentially increases the tension between the faculty and the administration, although it can also strengthen their ties and alignment behind a common mission. Accountability and assessment measures are intended to promote transparency and efficiency, but they potentially increase costs through more stringent regulations. Surely these criteria further challenge the prospects for inclusion, as participation is very time consuming. The college campus is a crucible of scrutiny.

Whenever you have multiple constituencies with conflicting demands, it is a challenge to balance them fairly. There are many reasons why this is so. People may bring strong emotional considerations, specific interests, or rigid points of view to bear on decisions that affect the commons. That is the very essence of politics. Democracy intends to provide an arena that turns differences into policy. When the process is successful, it is usually the result of layers and sequences of compromise. When the process fails, people are unable to see beyond their own interests, or they are unwilling to compromise. Good governance is easily subverted, too, by unilateral action disguised as consensus.

Often this is a perceptual issue. People may not directly grasp the relationship between their individual needs and the common good. It is easy to confuse one for the other and then obfuscate this mistake with elaborate rationalizations. Sometimes it is a problem of scale. People have perspectives that may be reasonable within a tightly bounded framework but entirely unhelpful in another context. Or they may not grasp the cascading ramifications of a seemingly sound solution. This can work two ways, as local decisions or points of view falsely generalized, or as global decisions that don't sufficiently account for local variations. Lastly, good governance may be elusive for reasons of intention. If only one party seeks to subvert the process through selfish behavior, using the system for exclusively private ends, an enormous amount of effort and energy goes into counteracting that behavior.

These are the basic realities of organizational life. We expect good governance, but we are not always equipped to provide it. We aspire to participate, but we may not know how to do so. We are quick to point out what is wrong with a system, or to provide ready solutions, only to learn that not everybody agrees with us, and the reality of implementation is subtle and nuanced. Administrative leaders know these challenges well, as they sit at the organization's center, striving to move it forward, balancing the expectations and aspirations of the constituents, with the realities of implementation.

Whether one is a president, a vice president, a provost, a dean, or a sustainability manager, in a system that is small, medium, or large, or whatever one's leadership function, one is compelled to follow the criteria of good governance while demonstrating accountability, providing a compelling vision, and adapting to the changing circumstances of a volatile economy in what amounts to new terrain for higher education. Good governance requires wise leadership. It demands a sophisticated understanding of how to implement change, while satisfying multiple constituencies, and building collaborative efforts. More than ever, good governance is a formidable challenge. Yet there is no alternative. If colleges can't be showpieces for participatory governance, how can we expect to educate students as to its virtues?

Leadership and the Sustainability Ethos

"Sustainability" and "ethos" meet on the terrain of good governance. "Sustainability" conveys ecological awareness (broadly considered), and "ethos" suggests values, morals, character, and civic engagement. "Governance" describes the organizational and political processes that allow a community to make and implement decisions. "Good governance" establishes the specific criteria that inform the decision-making process. How might a sustainability ethos influence how a campus approaches governance?

The sustainability ethos is a world view derived from evolutionary ecology and from earth-systems sciences. We live in a planetary environment, often referred to as the Anthropocene, in which human action profoundly changes the ecological, evolutionary, geological, oceanic, and atmospheric dynamics of the earth system.[2] The sustainability ethos is a response to these profound changes. It offers an interpretation of human agency that recognizes our ability to alter the biosphere, respects the complexity of our ecological and evolutionary heritage, reduces the effects of human actions, integrates humans with the ecological landscape, and elaborates the relationship between individual behavior and ecosystem health. This response requires an ethos that allows individuals and communities to establish sustainable practices *and* moral behaviors that guide civic engagement and good governance. We superimpose moral layers, informed by sustainability criteria, on the interpretation of good governance. How do these moral layers become apparent?

First, they become apparent by lending an environmental urgency to campus governance. Rapid global environmental change is a contemporary reality, whether it affects the campus through climate change, through loss of biodiversity, through environmental pollution, through the provision of energy and food, or through the effects of a "superstorm." The campus has a moral obligation to raise community awareness about these issues because they will ultimately affect the lives of all of its students. Whether it takes the form of the immediate necessity for emergency management planning, the longer-term strategies of climate action planning, or the educational philosophy that guides curricular decisions, these are necessary priorities demanding the attention of leadership. A better understanding of the human condition in the biosphere is a necessary foundation for civic engagement.

Second, they become apparent by calling attention to the relationship between sustainability initiatives, organizational processes, and community behavior. As sustainability challenges become a campus priority, new obligations emerge, requiring moral consistency and reflective practice. The case for sustainability must be clearly stated, collaborative efforts are essential, measures of accountability and affordability are intrinsic to those efforts, and they are mobilized with participation and consensus. Sustainability criteria can serve as an ecological measure of organizational change.

Third, they become apparent by emphasizing how moral debates are intrinsic to a rigorous education. Since the late 1950s, college campuses have been sites of political protests, have generated many of the ideas that have spawned such protests, and have been platforms for social experimentation. Controversy and change is inevitable in the midst of interesting learning environments. Many faculty members use their classes to explore ideas, students have the freedom to construct new social networks and identities, and administrators get to experiment with various approaches to educational and infrastructure planning. As centers of research, campuses generate possibilities and solutions for their fields of study. The most dynamic institutions combine all these activities.

Fourth, they become apparent by linking principles to commitments. In thriving learning environments, students and teachers alike are concerned to practice what they preach. If they believe deeply enough in what they are learning, they will hold their institutions accountable to those beliefs. You can't teach civil rights, social justice, gender equality, or sustainability without aspiring to uphold the principles that follow. That's precisely why campuses can be such controversial places. The curriculum itself is subject to intense scrutiny, inevitably upholding and promoting a matrix of values and practices.

What are the implications for campus leadership? In view of these moral layers (urgency, consistency, controversy, and commitment), the role of leadership is to guide a campus through an exciting (and perhaps daunting) process of change, culminating in nothing less than a campus transformation. Awareness of the trials and tribulations of higher education, and of the pressures on campus leadership, further raises the stakes of this challenge. It requires a willingness to lead change, prompted by a strong moral commitment to the sustainability ethos, and an understanding that the campus is an ideal place to launch such an effort.

Yet an emphasis on sustainability presents additional governance demands for a college or a university. The leadership has to deal with the long list of accountability issues cited earlier, while compelled to move forward on climate action planning and other sustainability measures. Whether driven from within (the values and mission of the institution) or from without (the growing awareness of the necessity of sustainable solutions), or from both simultaneously, it is crucial to recognize that we now have another layer of difficult challenges and expectations for the campus.

Change and Transformation

Visit just about any college or university campus in North America and you will find an active interest in sustainability. In some cases, transformational change is already underway.[3] Since 2005, we have seen the emergence of sustainability as a comprehensive campus movement. Its relatively heterogeneous origins are stunning. Champions might come from any corner of the institution—facilities directors, chief financial officers, business school deans, the chief executive or the board of trustees, as well as the typically diverse assortment of students, faculty members, and staff members.

One of the most exciting recent developments in higher education (and beyond) is the number of new "sustainability officer" positions. These range from mid-level manager posts that may report to administrators in facilities, finance, academics, or operations. In smaller institutions the sustainability responsibilities may become part of a faculty member's portfolio. These sustainability professionals represent a vanguard of institutional change.

Yet these positions are all relatively new on campus (mainly developed since 2005), contain highly ambitious portfolios, and ask the individuals involved to do the work of a dozen people. Further, most of these individuals have some technical expertise, or they are sustainability generalists, but they are unlikely to have management or leadership training. In view of the complexity of the challenges, it is easy to become lost in the morass of institutional bureaucracy. Many of these sustainability professionals are frustrated that the pace of campus change is insufficient, given the magnitude of the planetary challenge. They also worry that their home institutions are not moving quickly enough, or that not enough of their colleagues understand the significance or importance of sustainability.

Three patterns stand out. Sustainability professionals want to better understand how to manage, facilitate and inspire transformational change. They agree that if they could get everyone in their institution working more closely on sustainability, they are more likely to achieve that transformation. Just about everyone suggests, too, that when the senior leadership, especially the president and board of trustees are strongly supportive, sustainability initiatives will flourish. Aligning these three patterns—managing change, strategic coordination, and presidential leadership—is a recipe for success.

State-of-the-art leadership education emphasizes how to "manage" change. There are many texts on organizational change, ranging from scholarly research to popular "how to succeed in business" guides providing assorted tips, anecdotes, and wisdom. Despite the great variety of organizational settings, cultures, and missions, and all the idiosyncratic behaviors that follow, there are consistent rules of thumb that represent the essence of leading change. Essentially, one has to persuasively state the case for change, work with all layers of an organization, develop a sense of ownership, understand the cultural landscape, prepare for the unexpected, and address the human side of the challenges.[4]

Social behaviors require the most attention. No matter how collaborative or commons-oriented a culture might be, people will usually wonder whether a proposed change will be "good" or "bad" for them as individuals. Change implies uncertainty, unpredictability, and variation. Some individuals thrive in these circumstances. Others (the so-called change averse) find the prospect unnerving and intolerable. Leaders learn how to coordinate these responses, balancing the pace of change with finesse, delegating tasks on the basis of an assessment of who is most capable of managing change, and carefully explaining the reasoning behind all of it.

Some organizational scholars equate effective management of change with high social intelligence,[5] arguing that if you read people well, understand how to motivate and challenge them, anticipate their responses and behaviors, and recognize their strengths and weaknesses you will be more likely to choose, manage, and delegate your personnel wisely. The widespread medium of social networking ensures an even higher profile for individual and group communication. People must always be ready to assess the intentions of other people. Now we are even more ready to do so. However, the wise leader will balance impression with actions, intuition with deliberation, and persona with authenticity. Good governance, wise leadership, and social intelligence are dynamically interconnected.

Transformational change begins with highly motivated people who care so deeply about an issue that they are willing to commit time and energy to changing the organizations that most matter in their lives. Leadership often begins with an individual who influences a group. Leaders designate collaborators, who then affect the trajectory of an organization. Such leadership can emanate from almost anywhere in an organization. Although any one individual's influence is often proportionate to his or her relative rank and authority, and access to resources matters, to get anything accomplished it is necessary to build coalitions, networks, and policies.

Good leaders assess all of these factors, coordinate the resources appropriately, delegate them well, and explain how all involved can work together toward a collaborative goal. A leader identifies points of leverage, identifies important actors, and then sparks action through inspiration, encouragement, and support. Essentially, leaders move creative energy through a system. The leader aligns diverse networks and coordinates their movement, building momentum and flow around common goals. Flow implies a reciprocal momentum that reinforces creative efforts and collaborative work.[6] Creative leadership is wisely tempered, eloquently conceived, and steadfastly persistent, promoting an enduring resilience, and mobilizing institutional momentum. These qualities contribute to a deeper understanding of how to inspire and manage change. Multiple leaders managing change together can inspire a campus transformation.

The "nine elements" provide a conceptual foundation for using sustainability as motivation for transformational change. They demonstrate how all aspects of campus life are intrinsic to the change process. Governance coordinates the change process. Leaders manage governance. When all members of a campus community contribute to the change process, the result is participatory governance and collaborative leadership. The prospects for transformational change are significantly enhanced when people work together to make it happen.

Mission and Vision

Good leaders are meaning makers. They provide a rationale for change, an explanation of its necessity, and a sequence for how it will unfold. In any organization, people must understand why their work is important, how their role contributes to a larger purpose, and why they should be committed to the organization. This is especially true in mission-driven organizations, and higher education is largely mission driven.

"Vision" is a word commonly used to describe a long-term, strategic understanding of an organization's mission and purpose. People with "vision" have uncommon insight and an ability to broaden how an organization "sees" itself. In higher education, members of the senior leadership are expected to have a visionary perspective. The sustainability ethos contributes to this broadened view. The role of a leadership team is to articulate that view so that it is widely understood, easily grasped, and aligned with the meaning and purpose of work.

The challenges for the leadership are to build a team that embodies a collective vision, to translate that vision into action, and to empower the campus to endorse the vision as its own. In the case of a vision of sustainability, the process may stall, or even break down, at any of these points. Any individual can make a compelling or even charismatic case for an idea, but if the case isn't deeply embedded in the mission of the organization, or widely shared by its constituents, it will diminish in importance when the leader moves on. This is a major challenge for sustainability in higher education. On some campuses sustainability initiatives only last as long as their champions. But meaningful change requires a collaborative vision, disseminated throughout the system, empowered by numerous change agents, so that the vision permeates all aspects of campus life and all campus constituencies.

This is what I tried to accomplish at Unity College. I used every conceivable opportunity to remind the campus about the importance of a comprehensive sustainability vision. As an exercise in time management, I counted the number of weekly meetings I had to attend, including "direct reports," committees, public presentations, board meetings, introductions, luncheons and dinners, convocation, commencement, and various plenary opportunities. Another college president once told me that the position is the best chance one will ever have to be an educator. I took her words seriously, and I used the presidency of Unity to ubiquitously communicate the necessity of our sustainability mission. I encouraged all senior staff members—indeed, all members of the organization—to do the same. I intentionally cultivated those who championed sustainability. And I also worked closely with people for whom sustainability wasn't a priority, trying to find a reasonable common ground. After several years, sustainability initiatives (some of which I wasn't even aware of) emerged of their own accord. I knew then that the campus had taken the vision as its own.

It is essential to integrate campus political constituencies around a common sustainability mission. One way of doing this is to revitalize or initiate an inclusive Campus Master Planning process involving board members, students, community members, and all relevant interest groups. At a smaller institution this can be a highly personalized process; at a larger one it may require extensive use of communications media. In either case, there is no better way to inspire creative thinking about the future of a campus, the importance of sustainability, the vitality of the community, and the coordination of planning. The Master Plan, available to all members of the community, becomes a guide to future action. It will be effective only if it is broadly consensual, fiscally sound, and reasonably adaptive. In difficult economic times, these processes must incorporate sound financial planning and wise investment. That's why leadership must also demonstrate how the sustainability ethos enhances the institution's long-term financial stability—a topic that will be covered in the next chapter.

The Senior Leadership Team

Implementing a comprehensive sustainability vision starts with an institution's senior leadership team. Depending on the size and type of institution,the membership of this team will vary. Typically it consists of the president and the chief academic, financial, administrative, advancement, enrollment, and student life officers. When the senior campus leadership is aligned, an institution can make extraordinary progress on any priority. If sustainability is the foundation of transformational change, it is necessary that all members of the senior team lead the way.[7] Far too often, the senior team will have one or two champions of sustainability. When this occurs, there may be progress in several corners of a campus, but the sustainability initiatives will be limited in their scope and effectiveness. This is the frustration that is typically expressed by campus sustainability managers. They perceive one or two senior leaders as allies, but they can't find a way to get all of the senior leadership on board. The leader who can most easily rectify this situation is the president.

An effective way to deal with this situation, regardless of institutional scale, is to have the sustainability manager (and his or her team) report directly to the president of the institution, or to serve on the cabinet. At a very large university, the sustainability officer should have the authority of a vice president. At a smaller college, there is every reason for the manager (or whatever the title) to report directly to the president. This enhances the stature, effectiveness, and visibility of sustainability initiatives. It ensures that the senior leadership is well informed about the status of various projects, including reporting protocols and partnership efforts. It helps the senior team understand how sustainability initiatives are financially viable, contribute to philanthropy, improve student life (including retention and recruitment), inspire programmatic innovations, and improve campus infrastructure. It provides a "conscience" for maintaining sustainability as a priority and thereby sends a powerful message to the campus community.

At Unity College, the sustainability manager reported to me. I emphasized four priorities:

• Mobilize the students and employ them as your work force.

• Make sure that the Climate Action Plan (as required by the American College & University Presidents' Climate Commitment) is moving forward.

• Work with students and faculty members on the STARS (Sustainability Tracking and Rating Assessment System) process.

• Network well with your peers in the region to ensure collaborative problem solving and mutual support.

By getting weekly reports on all of these matters from the sustainability manager, I became informed, supportive, and engaged.

The senior leadership team (including the president) can also find ways to build sustainability initiatives into job descriptions. In this way, a sustainability ethos can permeate every nook and cranny of an institution, from operations and procurement to residential life. The process for doing this will vary according to the size of the institution, and in some cases it might be best to pilot this idea in the department that's most prepared to succeed or can have the most visibility. This also worked well at Unity. After several months, members of the staff revised their job responsibilities and their supervisory approaches and generated a great deal of momentum for our efforts. The student life staff embraced this approach, setting up recycling competitions, poster projects, and other campus-wide community efforts.

One caveat: Beware of too much reliance on committees, a common mode of implementation. The intent of a committee is to ensure diverse participation, deliberation, and unified action. Sometimes committees work effectively. If they do, it is usually because their participants have sufficient power on the campus and when their membership includes campus opinion leaders. Too often, especially in the case of sustainability, committees are designated as a fallback position. They provide a platform for discussion, but they lack authority, autonomy, and agency. Or they are filled with people who care a great deal about a subject, but have so many responsibilities that they don't have sufficient time to move the committee's recommendations forward. Committees work best when they have the authority to implement their recommendations, and when they do so in conjunction with the priorities of the senior leadership team.

Community Networks and Partnerships

Sustainability initiatives are more likely to succeed when they are linked to regional and national networks. Community partnerships, corporate relationships, and participation with effective professional networks allow a campus to share its expertise, justify its efforts, view its work in a broader perspective, participate in collaborative research projects, gain visibility for its sustainability initiatives, and cultivate potential donors. An institution will choose partnerships and affiliations on the basis of its specific interests.

At Unity College, I wanted the campus to understand how our emerging sustainability vision had currency beyond our institution. Knowing we were a collaborative voice in a national movement inspired and motivated our work. We enthusiastically participated in many national sustainability partnerships. We signed the American College & University Presidents' Climate Commitment because I wanted to set rigorous standards for our climate action planning process. It was enormously helpful to work with a national group of committed presidents who were sharing their experiences, accomplishments, and challenges. The ACUPCC also allowed me to challenge the campus. As a member of the ACUPCC Steering Committee, I felt obliged to publicly substantiate our commitment, and to speak on behalf of it. We were a charter signatory of the Billion Dollar Green Challenge, a consortium of colleges with green revolving loan funds. (See chapter 5.)

We rewarded our campus innovators by sending them to a variety of conferences, webinars, and colloquia, encouraging them to offer presentations, join committees, and apply for awards. Our innovators included students, staffers, and faculty members. I encouraged them to collaborate with colleagues at other institutions. In my role as the college's president, I would attend many of these meetings, bring assorted teams with me, and provide support for the necessary memberships and collaborations.

We also supported as many community projects as possible. We worked with the locally based organization Veggies for All to develop campus gardens in partnership with the local food pantry. One faculty member[8] designed a community wind assessment program. His students would work with local communities to determine the feasibility of wind power. Several faculty members, in partnership with the Maine State Energy Office, organized our students to provide energy retrofits for low-income residential dwellings. Another faculty member founded *Hawk & Handsaw: The Journal of Creative Sustainability*. (See chapter 9.)

Whenever we could, we would host regional sustainability meetings on our campus. When there were regional or national students conferences, we would organize transportation for all neighboring campuses. We encouraged our students to present posters and presentations in community settings. We set up dozens of community service projects.

Of course I am proud of these examples, and grateful to the Unity community for their extraordinary efforts. Combine these projects with thousands of others taking place at hundreds of colleges and universities and you get a sense for the scope of what is possible. These community networks and partnerships are the ground floor of lofty visions and ideals. Such projects generate momentum, cultivate community trust, and support the sustainability ethos well beyond the boundaries of a campus.

Innovation and Accomplishment

Whenever I met a new member of the faculty or the staff, I would suggest an overriding priority. "Your job," I said, "is to train a new generation of sustainability leaders who can enter the workforce and make a real difference in their communities. Take them into the field, provide them with rigorous hands-on experiences, try bold and engaging teaching techniques, and deepen their enthusiasm for learning. If you come up with interesting new ideas and approaches, we'll find ways to support, encourage, and reward you." I asked the faculty as a group to overhaul the curriculum, and to develop programs that would build on our strengths, emphasizing science, sustainability, and service. We created in the senior staff a spirit emphasizing campus service. Our administration aspired to make things easier, provide opportunities, reduce bureaucracy, and open channels of communication. We kept our doors open and tried to be available as much as possible. We encouraged staff professional development, especially around sustainability, change management, and organizational process.

Like many resource-strapped colleges, we were always vigilant regarding our budgets. Our challenge was finding ways to reward innovation without adversely affecting the bottom line. We developed two modest revenue streams for supporting sustainability initiatives. First, I kept a discretionary fund in the president's office. It was designated to support students, faculty members, and staffers who had innovative sustainability ideas. We sent mixed teams to conferences, gave small seed grants to interesting campus projects, and worked with all campus constituencies to figure out inexpensive ways to try out interesting ideas. Second, we made sure that the college's Advancement Office was geared toward finding support for sustainability initiatives. We received numerous small grants to enhance vegetable growing infrastructure, install a wood pellet heating system, perform a comprehensive energy analysis, promote community wind energy assessment, build a small barn, and support faculty research.

A primary function of senior leadership is to provide learning and working environments that enhance collaboration, facilitate creative thinking, encourage risk-taking, and celebrate success.

In view of the daunting tasks of organizational change, the planetary sustainability challenge, and the volatile economy, I sometimes wondered whether my actions would make any difference. Every change agent asks such questions, and often despairs in the enormity of the burden. Yet the process of change yields its own rewards, and one can take great pleasure in the simple satisfactions of a job well done, achieving a common goal, and the sequence of seemingly minor accomplishments that add up to something much greater.

The task of leadership is to celebrate these accomplishments at every opportunity. There are dozens of ways to do this, including rewarding individuals, calling attention to innovative ideas, and recognizing collective achievements. It is crucial to establish an atmosphere of community accomplishment. We used every means at hand to provide such support.

What I find is that over time an institution increases its accomplishments when it demonstrates how individual efforts contribute to the collective good. As I consider the five years I spent at Unity College, I am amazed at the number of sustainability initiatives that are now permanent aspects of college life. It never seemed that we were moving as fast as we should have been, and I was always aware that our progress was tentative, precarious, and based on sheer perseverance and collective will. As a leader, I would remind people of all the good work they were doing, but how much improvement they still needed to make. I don't think this feeling ever goes away, especially in regard to higher education, our planetary predicament, and the necessity of sustainable solutions. Indeed, we won't rest until we commonly acknowledge that sustainability must become the basis of how we learn to live, and hence the very foundation of higher education.

Words and Deeds

On a college campus, a community in which people explore values, ideals, and aspirations, it is crucial to practice what you teach. The various constituencies (students especially) are always ready to point out hypocrisy, inconsistency, deflection, ambiguity, or ambivalence. Indeed, an important leadership challenge is to be willing to listen to such criticism, while simultaneously acting respectfully and exercising legitimate authority. I found that on a small college campus (but this is certainly true at all institutions) the community observes its leader very carefully. Symbolic gestures and practical life habits are closely scrutinized.

When I arrived at Unity College, there wasn't a presidential residence. I suggested to the board of trustees and the whole campus community that we construct a LEED Platinum zero-carbon residence that could serve as the focal point of our future sustainability efforts. It was essential that we do this in a way that would draw attention to our efforts without ostentation, demonstrating our commitment to sustainability, but in a way that would be appropriate to the resource-strapped ethos of our campus. Hence, the residence would have to be small and reasonably priced and would have to be both private and public. My family would live there (and try to have a private life), but as stewards of a home that would be very public place. Thanks to the inspiration and vision of Tedd Benson of Bensonwood Homes, we built Unity House,[9] a brilliant solar home that became a landmark and a source of pride for the entire campus. We afforded it by using the president's housing allowance as the monthly payment, and by taking a modest chuck of capital from our precious plant fund.

There is much to say about the various challenges that we encountered in initiating and completing this project, including legitimate skepticism from some campus constituencies who questioned the priority of such expenditures. In my view Unity House was a testimony to the college's commitment to sustainability, allowed us to bring distinguished visitors to campus, was an inspiration for fund-raising efforts (including a "green" residence for students), brought outstanding publicity to the college, helped recruit my successor, and made a wonderful public statement about living what we teach. The entire campus eventually took great pride in Unity House.

This project was one of the most visible examples of dozens of "practice what you teach" campus efforts, including growing more food, serving more local food in the cafeteria, serving local food at catered events, experimenting with a variety of alternative energy sources (wood pellets, wind), planting wildflower gardens, exploring composting and recycling options, rural car-pooling efforts, and various operational and facilities improvements.

Every institution has an array of choices along these lines. There is no question that such choices start at the top. When a college's president is a role model for sustainability, it is more likely that the whole college will join the effort, and the entire community will be better prepared to exercise sustainability leadership.

Despite our best efforts, we inevitably made mistakes, and there were countless ways that we could have done more. Further, the idiosyncrasies and inconsistencies of our efforts were inevitable. It is far better to use such mistakes or dilemmas as teaching opportunities than to be defensive about them or to fall into a precious political correctness that will only encourage detractors.

Temperament and Disposition

Reflective leadership demands a state of constant attention, observation, and personal awareness. To reconcile words and deeds requires a rigorous, constructive process of self-assessment. Am I consistent with my objectives? Am I fair and balanced in my assessments of others? Am I providing challenging and nourishing supervision? Do I fully understand the needs of this campus? Am I representing the college well? Am I stewarding its resources responsibly? Am I serving others well? I could easily fill a dozen pages with a long list of similar questions.

Leadership is a social behavior. You learn the vital importance of personal relationships. And you learn how to maximize those relationships to stimulate creativity, productivity, purpose, and morale. Understanding the ever-changing dynamic of those relationships is a powerful foundation for effective leadership. To effectively steer transformational change, you have to understand an organization's capacity for change while assessing how your own temperament and disposition influences that capacity. That's the essence of reflective leadership—knowing yourself in relationship to the organization.

The sustainability ethos is concerned with maximizing human flourishing in the biosphere. This maxim became my personal guide for reflective assessment. Am I doing all that I can to maximize human flourishing on this campus? My assumption was that I could do so by paying close attention to the quality of social relationships, and by calling attention to how sustainability initiatives could improve the quality of campus life.

My leadership philosophy came from an aspiration to merge these criteria into a seamless process of organizational change.

At Unity, I spent much time learning on the job. I often improvised, striving to assess each situation as it informed the whole picture. Many late evenings (and early mornings) I would reflect on the what, the why, and the how. I realized that to accomplish my leadership goals, I had to better understand the temperament and disposition of the campus, the waves and cycles of campus morale, our accomplishments and frustrations, and take them on as if they were my own, while also remaining detached and objective. Similarly, I had to find balance and equanimity within myself, to serve as the anchor and ballast for a campus experiencing a great deal of change.

The accompanying tensions were both public and private. I made them public by discussing them openly. I thought that an open and honest appraisal of the change process would be healthy and beneficial, serving as a way to further model the sustainability ethos. What is the point of transformational change if we're all unhappy, if we're not flourishing? How can we find a psychological center for the campus?

Yet this was also very much a personal and private challenge. As the chief executive, I was the public face of the institution. I carried its identity, vision, hopes and dreams, stewardship, and public persona. I was the lead educator, chief administrator, delegator, fiscal agent, arbiter of justice, diplomat, advocate, emissary, and negotiator. No matter how well these various roles are distributed or how much incumbent support the president summons, the responsibilities of these roles are always evident. People looked to me for guidance, consistency, wisdom, and compassion. But I could only attempt to find those qualities through contemplation and deep reflection. This was easily my biggest challenge as a college president, navigating the psychological tensions of leadership.

I have "named" these tensions as urgency and patience, boldness and compassion, innovation and tradition, autonomy and authority. In writing about them I am hoping they are evocative for anyone who takes a leadership role, in whatever capacity, or regardless of institutional scale. We become better leaders when we share the tensions we encounter and when we empathize with those who lead. While doing so we are more likely to enhance the sustainability of the leadership function, both personally and for the sake of the institution.

Urgency and Patience

For the entirety of my professional life, ever since I entered the environmental field (in the early 1970s), I have been motivated by a sense of urgency. Whether it was the Union of Concerned Scientists' famous minutes-to-midnight annual proclamation, or the shadow of rapid species extinction, or any of the ecological and human planetary challenges we are so familiar with, I have been exposed to relentless proclamations and warnings about the fate of the earth. The same time pressure pervades the sustainability ethos. Forty years later, I still feel this powerful need to act.

At Unity College that urgency was amplified by the changing circumstances of higher education. We faced the typical issues: declining infrastructure, the need for curricular revitalization, better marketing and fund raising, spiraling tuition costs, and better salaries and benefits. It was clear to our leadership team and the board of trustees that our college required substantial change. We were one of hundreds of small, tuition-driven, independent colleges that faced issues of survival in the highly competitive environment of higher education. We had to further distinguish ourselves in every conceivable way. It is not too strong to suggest that the college had to adapt to survive.

All the prerequisites were in place for an urgent change agenda. Yet despite this common understanding, there were still many people on campus who were change averse, and as the newcomer, I had to endure considerable skepticism regarding my ability to lead the change. There were quite a few people, especially longer-term faculty and staff members who told me bluntly they had seen presidents come and go and they didn't think the college had the capacity to change in the ways it had to change. I know from speaking with other college presidents that my situation was not unique.

I understood that it would be easiest to work with staffers and faculty members who were most ready to change, and difficult to work with the change averse. I also had to recognize the importance of emotional modulation, both for the campus and myself. If I moved too quickly, only a few people would follow. Too much urgency can be fatiguing and the cause of burnout. If I pushed too hard, I would overtax all of us. If I moved too slowly, the college wouldn't change as quickly as it should. Too much stability can lead to inertia and laziness. If I was lax, I would be perceived as ineffective, and I would be personally unhappy.

Every institution and every leadership team has to determine the proper balance between urgency and patience. It can do this best by regularly considering morale indicators. Similarly, senior leadership can create a positive attitude about constructive change by periodically celebrating and attributing tangible accomplishments, rewarding the innovators, and helping promote an attitude that institutional change is meaningful, enjoyable, and improves communication.

The sustainability agenda was ideal for conveying the urgency of change. Its values were conducive to campus identity. Its implementation allowed us to become a distinctive regional and national voice. The "assessment" requirements, such as the ACUPCC climate action planning process and the STARS program, allowed for a convergence between campus and planetary urgency. Still, our resources were limited and it took discretion and deliberation to build a consensual approach to this challenge.

I considered the balance between urgency and patience on a daily basis. I explored this issue with my senior team and in other campus settings. As a leadership team and as a campus, we considered the pace of change. How quickly should we proceed? What is the campus carrying capacity for change? There is no formula for making this assessment beyond common sense, broad consultation, and social intelligence. We promoted a change agenda to the limit of people's capacity to internalize the process, erring on the side of urgency, but never forgetting the need for patience. I am convinced that raising this issue publicly helped convey its importance and brought a deeper understanding of how we could collaboratively achieve the required balance.

Boldness and Compassion

Talking about change is the easy part. Getting it done is another matter entirely. Ask any politician! When you take bold action, the first step is to consult widely. Then you have to be sure that you're moving in a direction that will maximize your leverage, build social capital, and have tangible, short-term results, while keeping your eye on the long-term strategic vision. This requires both timing and finesse. If you consult for too long (see the section on urgency and patience), you may miss your moment; if you act too quickly, you may inappropriately shock the system. If the timing is right, and if your actions are considered significantly bold, you will send an important message to the campus: We really are doing this! If those actions are clearly explained, people will understand why the campus is following a particular path, and they will grasp the relationship between ideas, intention, and action. When that relationship is questioned, you lose the momentum for change.

Whenever you take bold action, there will be people who disagree with you for reasons of substance, values, timing, finance, or ego. I also learned to never discount the inertia of curmudgeons. I followed four general approaches. First, I made sure that all members of the senior team were in approximate agreement on our collaborative decisions and could effectively discuss the reasoning behind them. Second, I would send a note to the campus community explaining our approach. Third, I was respectful that those who disagreed needed a venue for doing so. Fourth, I never responded defensively to critics, bearing witness to their issues and listening carefully to their concerns without yielding on our direction. This is especially important in contentious situations. However, in almost every case I would have to set limits beyond which I challenged disrespectful behavior, typically in private but sometimes in public.

I found the work of the Buddhist nun Pema Chodron especially instructive in this regard. In *No Time to Lose*, she describes the importance of absorbing other people's stress as a means to find balance and equilibrium, as a way to look deeply into your own awareness, while building compassion for the suffering of others. This is useful advice for organizational leadership. By attempting to absorb the stress, by understanding its causes and anxiety, leadership acts to restore balance, build confidence, and promote an atmosphere of constructive change. Often, resistance to change is an indirect way to avoid stress and anxiety. If one recognizes this energy, it can be redirected accordingly.[10]

Innovation and Tradition

A classic response to an initiative for change at an educational institution is "We have never done things this way" or "This goes against the traditions of our campus." Such a response can come from any campus constituency—from faculty members who are hesitant to change the curriculum, finance folks who are resistant to opening the budget process, or alumni who have romantic images of the campus from many years ago. It is very difficult to overcome a romanticized past, overhaul the curricular canon, or change the way the campus conducts its financial affairs. These are just several examples of the ways in which tradition can be used an as excuse to avoid change. Opponents of change often romanticize tradition. They create stereotypes of a glorified past that has a questionable basis in reality. Similarly, it is easy to stereotype agents of change and trivialize their vision.

I view this as the checks and balances of innovation and tradition. There is a measure of truth to both stereotypes. There may well be important values and processes embedded in campus history, and there are change processes that may not be thoroughly conceived. The appropriate response is to find the balance between innovation and tradition so that they mutually reinforce one another.

What role does senior leadership serve in balancing these perspectives? Every institution has a set of values, accomplishments, and narratives that are deeply embedded in its identity. The president (and the team) is expected to further exemplify, embellish, and promote that identity. Their challenge is to find the elements within that identity that form the basis of new directions.

At Unity College, I tried to find the elements of campus traditions that were most important to all the constituents. You figure this out by consulting with the most senior members of the staff and the faculty, with alumni, and with members of the board of directors. Then I attempted to link those traditions to how we conceived of sustainability. At Unity, there was great pride in the number of students who valued outdoor experiences—hunting, fishing, search and rescue operations, ecological research, wildlife care, organic farming, enforcement of conservation laws, and so on. We had to figure out how to take the best of that work, use it as the basis of a new sustainability vision, and weave new ideas about the future of the college into the mix. Similarly, the college prided itself on its resource-strapped, grass-roots spirit. Hence we described our efforts as frugal sustainability, to indicate our low-cost, practical approach. We strove to incorporate tradition and innovation in the cultural framework of the campus.

Autonomy and Authority

When an entire campus is engaged in sustainability initiatives, various constituencies will require autonomy to make good decisions on their own. As they participate in local decision making, they will feel most empowered to implement those decisions if they have sufficient responsibility and authority. If their meetings and decisions lack authority or the necessary resources, or if they are repeatedly challenged and overturned by senior leadership, they will perceive their own decisions as less consequential. Building trust is a reciprocal process. When constituents are empowered by leadership, they are more likely to take their decisions seriously. This builds confidence in the governance process, builds respect for senior leadership, and ensures consistent engagement.

Yet in any governance system, delegating authority creates risks. There are loose cannons, subversive personalities, and cynical perspectives, and there is always some ambivalence toward authority. These challenges are independent of scale and are ubiquitous to governance processes. It only takes a few difficult controversies to unleash waves of distrust. I'm sure that most every college president or CEO confronts these challenges.

There are three important risks attached to "distributed" leadership. First, it is possible to relinquish too much control, which may result in disorganization and inefficiency. Second, too much delegation might imply an abnegation of responsibility. Third, you can claim that you're delegating authority when in reality you are making all of the most important decisions yourself.

If senior leadership consistently acts in alignment with its stated objectives, if that alignment includes building system-wide trust and confidence through delegated authority, if it demonstrates an understanding of how to meld responsibility and autonomy, and if it builds public discussion around these challenges, it is more likely to succeed. Engagement requires participation. Leadership must be clear as to the terms and responsibilities of this engagement. You gain respect when you clearly explain your objectives, your reasoning, and your expectations, involving the whole community in meaningful discussion, while having clear lines of authority and accountability.

The Sustainability Outbox

There are many ways to successfully implement a sustainability-oriented campus transformation. Some universities focus on academic change— building new degree programs, developing sustainability requirements across the curriculum, revitalizing the professional schools, merging departments and institutes, or starting brand new ones. Others are making great strides in developing sustainable campus infrastructures—achieving climate neutrality, building renewable-energy facilities, reducing the waste stream, or growing organic food. On some campuses, these processes are mutually reinforcing and enhancing.

These initiatives never happen of their own accord. They start with the inbox, the place where ideas and priorities gather and germinate. They are sifted and vetted through intentional, strategic, and comprehensive planning. They are modified and enhanced as they make there way through the real world challenges of everyday campus life. They become successful as more people are convinced of their importance and virtue. When an initiative is fully incorporated, it reaches the sustainability outbox.

A thoughtful and creative governance process is the foundation of every successful and enduring initiative. Every initiative has a champion, and every champion has taken a leadership role of some kind. Every champion also has a story to tell about what works and what doesn't, what brings people together and what drives them apart. These stories become the sustaining leadership narrative for an entire institution.

Transformational change is challenging, time consuming, controversial, and risky. It takes perseverance, balance, clarity, and inspiration. It will be challenged and criticized. However, consider the outcome. When a campus is engaged in a transformational process, it can dramatically influence how an entire region thinks, acts, and lives. The college or university fulfills its role as an educational leader. It generates new ideas, solutions, and prospects. The campus enhances the quality of life for everyone who comes into contact with the institution. The sustainability outbox is the combination and culmination of all of these good works, made possible by good governance and wise leadership.

5

Investment

What Is Sustainable Investment?

Similar to many other small colleges and universities, Unity College was just several poor enrollment cycles removed from severe austerity. Like many presidents, I worked closely with the senior team to improve our recruitment and retention efforts. We "tracked the numbers" weekly so as to be able to plan a realistic budget. When we were meeting our targets, everyone breathed a sigh of relief. If the numbers appeared to be down, we had to make some difficult budget decisions. It was always amazing to me to observe how the weekly admissions reports could affect the moods and confidence of an entire campus. Over time, we learned how to improve recruitment, increase retention, and build predictability into the admissions/budget/revenue cycle. Still, when the economic downturn hit in 2008, we did our share of scrambling. As our endowment was so small, we didn't lose nearly as much revenue as other institutions, but we worried that many of our students could no longer afford to come to (or stay in) college. Despite our best fund-raising efforts and our interest in generating alternative revenue sources, the reality for Unity (as for hundreds of other institutions) was evident: Tuition was our most important source of revenue. There was also great uncertainty on the expenditure side. We could anticipate some unexpected costs and budget accordingly. We could engage in some long-term planning to balance salaries, infrastructure, and other expenses. But there were two expenditures that we couldn't predict: energy costs and health-insurance premiums.

The trustees were an interesting assortment of regional businesspeople, environmental advocates, outdoor enthusiasts, and alumni. What they lacked in philanthropic depth, they made up for with heart and common sense. The trustees were dutifully concerned about the financial future of the college. They understood the various financial viability ratios that measure a college's fiscal health. They were hesitant to take on debt and reluctant to take investment risks. That didn't leave the college with much flexibility. Six months into the job, I thought: "OK, let's see, no significant philanthropy, low annual giving, modest tuition, no summer revenue, not much debt flexibility. Our salaries are below our peer institutions, we've got major infrastructure needs, and we could really use better equipment and facilities. What am I going to do?"

I wondered whether sustainability might provide a framework for tackling some of these issues. We could use the idea of sustainability to revitalize the curriculum (thus attracting and retaining students), save on energy costs, reduce health-insurance premiums (by emphasizing wellness), and launch a new approach to fund raising. The trustees heartily endorsed the idea of saving money on energy, were thrilled with the prospects of utilizing renewables, were delighted to support growing more food on campus, and were enthusiastic about the prospect of the college's becoming a regional incubator for sustainability businesses, assuming it would be a great way to contribute to the economic welfare of the community. It took a few years to fully implement these ideas. And it was a challenging process. I had to work closely with the board (and the senior team) to better understand how we could make this approach work. I knew I could make the educational and ecological case for sustainability, but what about the economic case? More important, could I demonstrate the relationship between all three approaches? The first conceptual step was to fully grasp how to link sustainability to traditional concepts of investment, capital, and endowments.

What is "sustainable" investment? From a strictly financial perspective, it suggests that the campus endowment will be more oriented toward ecologically and socially responsible equities. That campus budgeting will give priority to energy efficiency and conservation. Campus leadership will investigate and implement financial incentives to stimulate comprehensive, ecologically sound approaches to energy, materials, and food infrastructure. The campus will work with the regional community to develop sustainability markets, serving as a dynamic economic multiplier for sustainable businesses. More broadly, campus sustainable investment implies ways of reconsidering all of the institution's capital assets.

We often conceive of investment as a financial term, strictly defined as "the process of exchanging income during one period of time for an asset that is expected to produce earning in future periods."[1] However, in common parlance "investment" has a broader meaning. We use it to convey the amount of time we are willing to spend doing something in the hope of a future reward. As individuals, we think about the resources we are willing to commit to a project or process. These resources might include time, money, knowledge, talent, and effort—a range of abstracted or tangible qualities that coordinate our personal assets. When making a decision, or in thinking about the consequences of an action, young people especially are advised to invest wisely in their future. The cost-benefit analysis of an investment goes way beyond a simple financial assessment.

The concept of investment can be similarly expanded for institutions, especially colleges and universities. The endowment is the repository of the system's financial assets, reduced to an investment portfolio. The assets (time, money, knowledge, talent, effort) of a system transcend this financial reduction, although a financial equivalence is always presumed. When institutions engage in planning and thereby balance present needs with their anticipation of the future, they have comprehensive discussions about the deeper meaning of investment. On which projects should we spend time? In what ways will our present efforts be rewarded tomorrow? How do our knowledge contributions manifest themselves in future outcomes? How should we be spending our money?

In this chapter, I present a variety of approaches for thinking about sustainable investment. I begin by expanding the idea of capital to include social, ecological, and intellectual capital as well as the traditional concept of financial capital. When we consider the varieties of capital, a campus may have many more assets than it customarily assumes. Similarly, I'll expand the idea of an endowment to include an ecological portfolio of campus assets. If we consider the campus as a place in ecological *and* economic space and time, we can begin to approach the deeper meaning of sustainable investment. I'll go on to discuss two tangible ways to integrate ecological and economic scale: the climate action plan and ecological cost accounting. That will lead to a discussion of campus eco-finance, including green revolving loan funds. With targeted investment, a campus can serve as a sustainability capital incubator, launching markets, businesses, and thriving sustainability partnerships.

The Varieties of Capital

Like investment, capital is a surprisingly complex concept. A financial definition suggests that capital represents assets available for use in the production of further assets. What are assets? Anything of material value owned by a person or company. Kenneth Boulding, in *Ecodynamics*, one of the first ecological economics texts, defines capital as "the stocks or populations of all economically significant objects." Boulding suggests multiple forms of capital, including all the stocks of commodities, energy and potential energy, and knowledge. He further defines "human capital" as "the bodies and minds of the human populations insofar as they are economically relevant."[2]

Several prominent ecologists and ecological economists use the term *natural capital* to signify "the goods and services from nature which are essential to human life."[3] Robert Costanza elaborates:

Natural capital is the extension of the economic notion of capital (manufactured means of production) to environmental goods and services. A functional definition of capital in general is "a stock that yields a flow of valuable goods or services into the future." Natural capital is thus the stock of natural ecosystems that yields a flow of valuable ecosystem goods and services into the future.[4]

A good way to specify the meaning of human capital is to distinguish it from social capital. Lew Feldstein and Robert Putnam describe social capital as "the collective value of all 'social networks' (who people know) and the inclinations that arise from these networks to do things for each other ('norms of reciprocity')."[5] In more traditional economic terms, social capital consists of "stocks of social trust, norms and networks that people can draw upon to solve common problems."[6] Another approach to human capital is to emphasize *intellectual capital*, or the "collective knowledge of the individuals in an organization or society."[7] Some businesses, recognizing the importance of intellectual capital, have even developed ways to measure it, assuming that it is a quantifiable asset.[8]

In the next section, I'll discuss these varieties of capital through the perspective of sustainability. How can we assess, organize, project, and perhaps even measure sustainable investment by considering wealth and capital in these terms?

Campus Eco-Capital

When you consider this broadened approach to capital (natural, social, intellectual, and financial), you realize that college and university campuses have an extraordinary variety of assets. If you measure a campus's capital assets in exclusively financial terms, you overlook other forms of wealth that are additional prospects for sustainable investment.

Financial capital is the standard measure of a campus's financial health. A package of comprehensive ratios serves as the foundation of best-practices financial assessment—viability ratio, primary reserve ratio, and net income ratio as calculated through expendable net assets, plant debt, total revenues, total operating expenses, total non-operating expenses, and change in total net assets. Although there are different reporting mechanisms, and although there is some controversy associated with the extent to which any particular approach is best suited for higher education, all the reporting mechanisms are organized around cash flow, reserves, depreciation, income, and expenses.[9] These approaches assume that wealth is best measured through its representation as money.

What happens when we assess a campus's wealth by considering the varieties of capital? Although there have been several attempts to measure intellectual capital as a campus's asset, these mainly focus on standard academic productivity criteria—publications, collaborations, research grants, credits, courses, and other ways of verifying knowledge production. A similar approach could be used to organize, catalog, and understand campus sustainability initiatives—sustainability as knowledge production. The most conventional way to do this would be to use the proposed knowledge production as a base layer of measurement, and then apply knowledge production to all academic projects that are oriented around sustainability research, teaching, and service. A campus could ostensibly measure how much of the intellectual enterprise is devoted to sustainability, providing at least a qualitative assessment of its intellectual capital. This is an asset in the sense that it adds value to the institution's prestige and research, leveraging its potential to develop conceptual breakthroughs in the sustainability field.

Colleges and universities rely heavily on promoting social capital as a means to enrich campus life, stature, influence, and effectiveness. Campus life is typically rich with affiliations, clubs, networks, and associations. The desire to increase one's social capital is a primary reason for attending a college. Although it is hard to precisely measure social capital, or to comprehensively assess its importance, there may be a direct correlation between social capital and quality of life.[10] The sustainability movement promotes social capital by emphasizing community partnerships. All the sustainability initiatives on a campus, including food growing, recycling, energy efficiency, collaborative research efforts, and service projects, involve people working together to improve campus life. Pick up just about any college's catalog (or peruse just about any college's website) and you are likely to see photos of happy and engaged people working together on sustainability projects or in sustainability programs. The social capital accrued through campus sustainability projects adds (perhaps immeasurably) to a campus's wealth.

All campuses have natural capital assets. Natural capital is typically interpreted as the visual appeal of a campus's landscape ("Please visit our beautiful campus") or as a campus's exceptional location ("Our campus has easy access to the diverse cultural resources of our city"). Campuses are built environments in natural places. They may be "endowed" with energy assets (solar gain, wind, geothermal), ecological assets (habitat, farmland, wetlands, watercourses), wildlife, and all of the intrinsic values and services of the ecosystem. Ecological economists refer to these assets and functions as ecosystem services.[11] These include services related to provisioning (food, water, minerals, energy), regulating (carbon sequestration, waste decomposition, air and water purification, crop pollination), supporting (nutrient dispersal and cycling, seed dispersal, primary production), and culture (inspiration, recreation, scientific discovery). All campuses contribute to ecosystem services. These services can be assessed for their value, although doing so requires a feasible translation between economic valuation and ecological processes. This translation challenge is not embedded in most CFOs' job descriptions. Nevertheless, there are some interesting attempts to plan for campus ecosystem services, including projects at Yale University, Reed College, and the University of Georgia, although they mainly deal with landscape design, and only abstractly refer to economic valuation.

Translating the varieties of capital into conventional valuation mechanisms is beyond the scope of this book. What I wish to convey is that sustainable investment entails an understanding that wealth, assets, and capital come in many forms. There are diverse ways for a campus to invest in a sustainable future, and it can do so more effectively by broadening its understanding of value. Perhaps it is sufficient to suggest that sustainable investment should consider "campus eco-capital" as the financial, natural, social, and intellectual capacity of a college or university campus.

What Is an Endowment?

An institution's endowment is the primary instrument that enables it to organize its financial investments. The original meaning of "endow" is "to enrich, mainly through property." An endowment represents a financial legacy. Whereas Harvard University's endowment is about $32 billion, some small institutions have endowments of less than $5 million. Regardless of size, the endowment plays an important role in all aspects of the college's finances, from providing scholarships and faculty positions to supporting annual operating expenses. As such, the endowment provides an important source of revenue and its financial performance can make a big difference in the annual operating budget. This is especially true of smaller colleges and universities that have come to rely on this supplementary income. However, during the economic downturn of 2008–2010 even the institutions with large endowments were significantly affected by the loss of expected revenue.

An institution's endowment represents a tangible and symbolic narrative of its history. Most of the institutions with the largest endowments are elite colleges and universities with many famous and wealthy alumni and extraordinary records of accomplishment, prestige, and influence. Indeed, the size of an institution's endowment is in itself a measure of this prestige. Essentially, an institution's endowment, if sufficiently ample, allows the institution to operate for many decades, providing security, longevity, and legacy.

Colleges with small endowments place great emphasis on increasing the size of those endowments, both for the obvious financial benefits and so as to project confidence, stature, and prestige. A major endowment gift is an important symbolic vote of confidence in the future of the institution. Just after I left Unity (a small and young institution, founded in 1965), the college received an anonymous gift of $10 million, which effectively quadrupled its endowment.

Unrestricted endowment gifts are the most appreciated gifts, because they provide maximum financial leverage for investment or operating expenses. However, many endowment gifts are restricted for scholarships and positions, or for special projects, buildings, and initiatives. In these cases, the donor has a substantive wish, and the institution is then obligated to use the gift for specific purposes.

An institution's endowment has much more than financial meaning. Deeply embedded within its relevant restrictions lies a values statement about what the donors most care about. Although the endowment's managers are mainly obligated to ensure the effectiveness of their investment strategy, they may be scrutinized for the values embedded in those investment decisions. Like everything else on a campus, endowment investing is subject to careful scrutiny: Where is our institution's money going? What are we investing in? Does our endowment reflect our values?

Sustainable Endowments

The sustainability ethos can be applied to endowments. Why shouldn't the endowment operate as a tool for "green" investing? There is a growing movement to assess endowments based on ecological and social criteria, including divestment strategies.[12] In 2007, supported with a consortium of foundation funding, the Sustainable Endowments Institute introduced the College Sustainability Report Card, a tool for evaluating a campus's sustainability efforts as well as its endowment practices. The Report Card's grading system assesses policies and practices in nine main categories: administration, climate change and energy, food and recycling, "green" building, student involvement, transportation, endowment transparency, investment priorities, and stakeholder engagement.[13]

Let's look more carefully at the "investment priorities" and "shareholder involvement" sections. Here the Report Card is further specified to include renewable energy and sustainable investment, community investment, on-campus sustainability projects, donor fund options, optimizing investment return, proxy vote decisions, stakeholder involvement, school community input, and sustainability voting record. Three Report Card assessments are relevant here. "Campus sustainability initiatives outshine endowment sustainability activities, a significant percentage of schools have endowment investments in renewable energy funds, and schools are weakest in Shareholder Engagement and Endowment Transparency categories."[14]

Campuses are making inroads using the endowment as an investment tool. In all, the endowments of American colleges and universities amount to $300 billion (as of 2009). Between 2008 and 2009, the number declined from $413 billion to $325 billion. The "green investment" potential of this money should not be underestimated.

An Ecological Endowment

If we expand our concept of endowment beyond financial criteria, as we did with capital and investment, it takes on expanded meaning. Recall the definition of "endowment" as "enrichment through property." What is property, if not ecosystem services organized through ownership? Many campuses have significant endowments of property in the form of land, buildings, and goods. If the endowment calculations were to include an economic valuation of ecosystem services, how would that change how we assess campus wealth? If the campus were to perceive its habitation as a reciprocal relationship with the ecosystem, would it think about its investments differently?

A campus's ecological endowment is an assessment of its ecosystem services. With the financial endowment, there are guidelines that protect the drawing down of reserves. Indeed, some institutions have been harshly criticized for keeping too much of their capital in financial reserves when they could be spending more in the present. Prudent financial planning suggests that you spend only a small portion of the endowment. Similarly, the campus can better "insure" the viability of its ecosystem services by carefully assessing how its consumption behaviors affect its ecological endowment. In this case, there is an opposite dynamic. We tend to draw down ecosystem services to protect financial viability. This is a conceptual challenge for sustainable investment. It is complicated by the difficult measurement juxtaposition between ecological and economic valuation. Further, there are conceptual challenges in comparing ecological and economic measurements of time and space, and in understanding the relationship between local and global ecologies and economies.

A Campus in Ecological and Economic Space and Time

Imagine the college campus that you are most familiar with. Focus on the ecology and geography of the place. What is the topography like? What flora and fauna live there? Where are the watercourses? You can do this whether your campus is in a remote rural area or in the middle of an enormous city. You can extend this thought challenge by imagining the campus terrain before it was a college, then going further back in time and imagining the place as it looked before human settlement.

Refer to my earlier discussion of natural capital, ecosystem services, and ecological assets. Choose any of the assets and trace how they have changed over the last 500 years. If the landscape was heavily forested and is now a built environment, then the energy assets and perhaps the biodiversity assets were significantly altered. On the other hand, the transformed, built environment may now have better prospects for wind or solar energy installations. There are many ways you can trace the environmental changes over time. It is instructive to consider how those environmental changes transform, alter, augment, or diminish the ecosystem services.

The main challenge of this exercise is to consider the dynamics of ecological space and time as applied to our broadened conceptions of capital and investment. To assess the ecological endowment as a long-term investment requires an expanded perspective of the campus. In essence we are shifting time scales. Once those shifts are made, the ecological endowment takes on a new meaning. The campus undoubtedly has "valuable" natural capital assets.

Economic time and space functions at a different scale and uses different measurements. We typically consider financial investments within the span of several generations at most, and more typically in time frames of 5–10 years. The same is true for how we scale returns on investment, interest rates, debt, and all of the financial viability ratios described earlier. Money is the standard measurement of value in a market system. Its calibration requires the scale of economic time. This is exemplified by how we assign value to labor (hourly wage or annual salary). We measure a person's "net worth" by virtue of financial assets accrued over a lifetime. In the case of individuals, we attempt to expand the time scale of economic measurement through the concept of inheritance. In the case of communities, we do this through banks, corporations, or the state.

When applying financial measurement to space, we calibrate the value of square footage or acreage. Perceived ecosystem assets may influence these measurements (natural beauty or wealth in natural resources), but typically it has more to do with the perceived value of a location, its proximity to other locations of value, and the financial benefit of the location's production or commercial value. Ecosystem services transcend acreage boundaries and can't be reduced to property designations.

Is it possible to find places of convergence between economic and ecological scales of measurement? A sustainability ethos demands such a convergence. However, the onus for this challenge is to find suitable conversions from ecological to financial concepts. This is inevitably so in a market society. If the standard measure is money, then our challenge is to find ways to broaden our financial measurements so as to include ecological criteria. Can we accommodate ecological value within economic criteria? If so, what kinds of calibrations will be necessary? In the next few sections, I'll discuss climate action planning and ecological lifecycle analysis as interesting applications that navigate this conceptual juxtaposition.

Climate Action Planning

Understanding climate change requires a sophisticated knowledge of biospheric space and time. You have to grasp biogeochemical cycles, geological time scales, and the flow of oceanic and atmospheric circulations. It is very hard to link specific, local actions and behaviors to these biosphere processes. Emissions of greenhouse gases are a tangible and measurable concept in this deep scalar web. We know that greenhouse gases contribute to the amount of carbon in the atmosphere (parts per million). Thanks to climatological, geological, and paleoecological research, we understand that this ratio varies throughout geological time. However, we know that beyond certain limits we create biosphere-level instabilities that will dramatically affect oceanic and atmospheric circulations, in turn affecting ecosystems. In an attempt to prevent and mitigate unprecedented and potentially life-threatening consequences, we seek to limit emissions of greenhouse gases. To limit them, we have to measure them.

There are numerous guides and tools for calculating greenhouse-gas emissions.[15] Although this can get complicated, as there are many spatial and temporal variables, the basic measurement mechanism is accessible. You determine the metric tons of greenhouse gases that are emitted as a result of all campus energy-use and energy-production activities. There are direct and indirect emissions, and there are protocols that explain how these should be calculated. Attaining carbon neutrality, the ultimate goal of climate action planning, assumes that a campus achieves net zero carbon emissions. The carbon released is balanced through an equivalent amount that is sequestered or offset.

This carbon measurement is the basis for the convergence of ecological and economic scales. We wish to reduce carbon because we don't want to exceed specified atmospheric carbon ratios. We have derived these ratios from biosphere scale measurements, by comparing the amount of carbon in the atmosphere and in the oceans over geological-scale time frames. We can use financial measurements to assess the cost of reducing carbon emissions. This is the quantitative basis of the climate action plan.

The American College & University Presidents' Climate Commitment (ACUPCC) was organized to enable higher education to take a primary leadership role in climate action planning.[16] College and universities that sign the ACUPCC pledge to attain carbon neutrality as soon as it is feasible to do so. Feasibility is mainly assessed though financial readiness, although implementing the climate action plan (CAP) requires the close coordination of all nine elements of a sustainable campus. Financial measures play a prominent role in attaining climate neutrality. Can the campus afford this process?

Many campuses have used the CAP as a means to coordinate ecological and economic concepts through the language of sustainability. Almost all CAPs include comprehensive discussions of the institutional commitment to sustainability and refer to ecosystem concepts in their rationale. Indeed, many of the public versions of these plans serve as "branding" tools to demonstrate how the campus is implementing a variety of sustainability initiatives. They resemble investment portfolios in their description of the future of the campus. For two outstanding examples, see the CAPs of Arizona State University and Carleton College—beautifully designed, comprehensive discussions of sustainability, finance, and innovation, implicitly describing campus investment, and portraying the ecological endowment (although not employing that term).

These CAPs (and all the others) present the details of financial steps (economic measurement time scale) that will enable the campus to achieve carbon neutrality. The Carleton CAP has a Table of Values appendix listing 26 different aspects of the greenhouse-gas-reduction strategy, from energy-efficiency measures to behavior changes to new energy-production facilities. For every strategy, there is a corresponding column that includes estimated capital cost, average annual operating cost, average annual electricity savings, average annual gas savings, average annual energy savings, and net annual cost.[17]

Arizona State University's CAP has a section on financing that lists the various ways it will generate income to support its zero-carbon initiative, including "university-based, socially responsible investments; efficiency-capture opportunities; as well as ways to restructure traditional financing structures for more efficient, sustainable options." These new approaches to financing are intended "to develop greater collaboration between research, academic and operations of the university to increase our competitive advantage for public and private funding opportunities." The CAP then lists a suite of proposed funding sources and investment options.[18]

Arizona State University pledges to attain carbon neutrality by 2025. Carleton College hopes to accomplish this by 2050. A survey of all of the ACUPCC signatories shows many different target dates. When developing CAPs, many schools are hesitant to plan far into the future because they don't know what their financial situations may be, and because they can't predict the strategic goals of future administrations. Although those uncertainties are real, in my view they entirely miss the point of this project. What is brilliant about the CAP is that it demands long-term, biosphere-scale thinking and planning. The planetary challenge of climate change will not be solved if we are limited to traditional, short-term financial measurements. By using a time frame of 25–50 years, or even 75–100 years, an institution can broaden its conception of investment to include the varieties of campus eco-capital. The CAP does this by integrating biosphere-level thinking with financial cost accounting. It represents an interesting first experiment in the juxtaposition of economic and ecological scale thinking.

Ecological Cost Accounting

Ecological cost accounting attempts to bring environmental considerations into financial measurement techniques. The basic concept is appealing and accessible. Every material object has an ecological footprint.

Ecological cost accounting attempts to trace, uncover, and incorporate that footprint into a pricing scheme.

In chapter 3, I described the challenges of tracing the ecological/economic pathway of the Magic Marker. The production chain is complex, from the origins of materials in the earth, to the chemical transformation of those materials, to the energy used, to the resulting waste, and to the factory conditions at the production site. Similarly, the distribution process entails a global transportation network with many hidden ecological costs. Consumption involves the final breakdown of the product after its use. Ecological cost accounting attempts to scrutinize this entire sequence, examining all the links in the chain of production, distribution, and consumption and then assigning environmental and economic cost criteria to every link in the chain. True ecological cost accounting will translate environmental impacts to financial criteria at every step. Although this is a simple concept, it is very hard to implement.

There have been many attempts to do this, especially in the business community. In their chapter on accounting and finance in *The Green to Gold Business Playbook*, Daniel Esty and P. J. Simmons point out that this is still an emerging approach to assessing environmental risks, benefits, and liabilities. "The good news," they write, "is that the dynamic field of environmental accounting boasts a diverse array of tools for different objectives. The bad news is that this field is filled with a dizzying array of terms and definitions that are sometimes contradictory, given the lack of standardizations and relative newness of the field." In view of these uncertainties, they suggest that you "work with your sustainability team to put the best tools into practice."[19]

"Cradle to Cradle" design, is one of the best-known ecological accounting techniques.[20] Life-cycle assessment is based on "compiling an inventory of relevant energy and material inputs and environmental releases." It can get very complex, requiring sophisticated software packages. The Wikipedia article on life-cycle assessment lists numerous "variants" of the methodology[21]: cradle-to-grave, cradle-to-gate, cradle to cradle, gate to gate, well to wheel, economic input–output life cycle assessment, ecological-based LCA, and life cycle energy analysis.

Despite the seemingly daunting array of methodologies, I find all these choices inspiring. Thousands of business, institutions, campuses, and agencies are trying to figure out how to better integrate economic measurement and ecological scale. Eventually these systems will find common criteria and a reasonable standardization process, despite the difficulty of measuring some of the ecological impacts. For present purposes, the issue is how campus finances and accounting can reasonably accommodate these approaches. Before addressing that, I'd like to mention one business that is attempting to incorporate a version of these methods as a paradigmatic shift in its operational planning. Ray Anderson describes the philosophy, methodology, and implementation of the Interface Model in his book *Mid-Course Correction*. In a chapter on "The Prototypical Company of the Twenty-First Century," he describes the epistemology of the Interface Model, using a series of ecological/economic diagrams to illustrate integrated approaches to zero waste, benign emissions, renewable energy, closing the loop, resource efficient transportation. Anderson proposes a comprehensive redesign of commerce. His highly idealized diagrams include many concepts of sustainable investment. Anderson's objective was to design a true "natural capitalism."[22]

The challenge for college and university campuses is how to adapt aspects of these models for both operational and instructional purposes. How might such processes be initiated?

Campus Eco-Finance

By "campus eco-finance" I mean all the standard and alternative approaches for including ecological considerations in financial decisions. The first step for doing so is to find ways to quantify and measure environmental impacts. For businesses, Esty and Simmons suggest using seven steps:

1. Get Familiar with Common Indicators of Business Environmental Impact Metrics

2. Plot Your Company's Carbon Footprint

3. Measure Your Company's Waste

4. Assess Your Company's Water Footprint and Related Impacts

5. Evaluate Your Company's Impacts on Ecosystems and Biodiversity

6. Undertake a Life Cycle Assessment of Your Products and Services

7. Redo Each of the Analyses at a More Advanced Level[23]

Colleges and universities that engage with the Association for the Advancement of Sustainability in Higher Education's Sustainability Tracking and Rating System (STARS) already have a great foundation for this work. STARS requires equivalent assessments in each of these categories, then goes into significant detail as to whether a campus meets sustainability criteria.[24] The challenge for a campus's administrators is to work with the sustainability team to coordinate the financial and environmental metrics. All the procedures recommended by Esty and Simmons, and most of the STARS methodology, can be linked to traditional financial cost accounting. Every initiative has a projected cost, and many may result in cost savings.

After measuring environmental impacts, a second step is to gain familiarity with "best-practices" sustainable financial approaches, including understanding micro-finance and micro-insurance, using small loans, using private equity and venture capital for sustainability initiatives, using new capital markets (renewable energy, forestry, and carbon) for project financing, and using revolving loan funds. These are the topics taught in some of the interesting new "green" MBA programs.[25]

A third step is to explore a new conceptual foundation for campus finance by implementing a broader view of capital assets and the endowment. Campus leadership can initiate strategic and master planning processes that incorporate the varieties of campus eco-capital in their projections. Traditional SWOT (Strengths, Weaknesses, Opportunities, and Threats) analyses—the "best practices" of strategic planning—can easily accommodate comprehensive campus-wide discussions of intellectual, social, and natural capital. These are qualitative capital assessments that can inform strategic financial planning.

A fourth step is to initiate campus-wide academic and operational efforts to research, organize, and apply "natural capital" concepts to both campus master planning and strategic financial planning. There are innovative approaches for doing so. Of special note is the Natural Capital Project, which is "developing tools for quantifying the values of natural capital in clear, credible and practical ways." The Natural Capital Project's website refers to INVEST (Integrated Valuation of Environmental Services and Tradeoffs), a software program that uses geographic-information-systems software. Much of the Natural Capital Project's preliminary research is covered in the book *Natural Capital: Theory and Practice of Mapping Ecosystem Services*.

Perhaps these approaches seem impractical. Campus finance officials, dealing with all the financial challenges to higher education, are already stretched thin. Many campuses are hesitant to undertake climate action plans because they don't have the expertise or time to compile the necessary data. How can they be expected to implement campus eco-finance? They may require additional training, support, and cooperation. This is an area where there should be major interdisciplinary efforts by the sustainability, environmental, ecological, economic, and business programs on a campus. These programs should be coordinated with campus administration and applied directly to campus operations and procedures. Implementing a sustainability ethos requires close cooperation between business schools and sustainability programs. Many of the new "green" MBA programs are moving in this direction. Business students and faculty members can apply their work in campus settings.

Modular Budgets for Multiple Scales

Implementing campus eco-finance involves many options. These will depend on the size of the institution, the available personnel, the willingness of campus leadership to experiment, the kind of academic programs on campus, and the interests of the students, the staff, and the faculty. In view of these variables and the complexity of managing campus finances, it is useful to think of *modular* budgets. Consider these modules as "just in time" applications that can be adapted as circumstances require them.

For example, an ecosystem service analysis of a small patch of open space on a campus can be performed. Once its ecological value has been translated into a financial measure, it can be added to the endowment as a natural capital asset in an "ecological endowment" appendix. As the campus improves its capacity to implement ecosystem services methodology, a more comprehensive ecological endowment can be developed as a parallel and supplement to the traditional endowment. Eventually this method can compare the ecological and economic strategic values of various approaches to land use. Is a certain plot better left as open space, or should it become the site of a new LEED science building? Who will do this work? Students from the business school and the ecology department can work together under the supervision of the CFO and a faculty member.

Climate action planning functions similarly. The greenhouse-gas inventory is a modular budget linked to strategic energy-efficiency planning.

In small steps, it can be integrated with all campus financial operations. Ecological life-cycle accounting can become sequenced through a series of exploratory procurement initiatives. Starting with one important campus commodity (let's say paper), a Cradle to Cradle assessment can inform a sustainable procurement policy. As the campus becomes more familiar and comfortable with this approach, it can be applied to all procurement decisions. Hundreds of campuses are initiating variations of these approaches.[26] However, because these activities seem daunting or impractical, many campuses don't engage in them. Yet they potentially yield long-term ecological and financial advantages. Start with simple projects and work with what you have.

These are exploratory approaches. They require a spirit of experimentation, innovation, and creativity. They will involve false starts. They may be criticized by traditionalists. Almost all of the methods cited in this chapter require interdisciplinary research and interdepartmental cooperation.. We are still learning about how they work, in what ways they are helpful, and how we can best integrate ecological value with financial measures. Twenty years from now we will have an entirely new set of methods, operational tools, software packages, and epistemological foundations. Colleges and universities are well equipped (with intellectual capital) to contribute to this new body of ideas.

Revolving Loan Funds

The revolving loan fund is an exceptional campus eco-finance investment tool. It integrates many of the concepts of sustainable investments covered in this chapter, provides a means of financing climate action planning and procurement initiatives, and now has a robust case-study literature.[27] The basic concept is simple. Campus sustainability projects gain their financing from institutional capital. The money for these funds comes from a variety of sources—administrative budgets, student "green fees," endowments, utility companies, and donations. Essentially these funds are used to finance a variety of sustainability projects on campus. Projects range in scope from quick-payback investments, such as upgrading lighting fixtures in a residence hall or installing composting equipment, to more systemic projects, such as comprehensive energy retrofits or investments in alternative energy sources. The savings from these projects are then returned to the loan fund, and the money then can be reinvested in additional projects. The main challenge for a campus is to determine where the start-up capital will come from, and then to develop the procedures and protocols for determining how the money be used and reinvested.

Since the 1990s, more than fifty colleges and universities have established funds to finance campus-based sustainability initiatives. The California Institute of Technology has a Conservation Investment Program, Harvard University has a Green Loan Fund, and Stanford University has a suite of programs. These fifty schools include well-endowed institutions with copious resources, state colleges and universities whose budgets are approved by legislatures, and smaller, resource-strapped colleges whose revenue is derived almost exclusively from enrollments.[28]

Schools have reported average payback periods ranging from one to ten years, with a median of four years. After these loans are repaid, additional savings accrue to the school's operating budget. College executives, most particularly chief financial officers and members of finance committees, will be pleased to know that Green Revolving Funds (GRFs) provide reliable returns on investment ranging from 29 to 47 percent with a median of 32 percent. Even the most sophisticated endowment managers will have a hard time matching those figures.

The Billion Dollar Green Challenge is an effort to maximize these efficiencies and innovations through a nationwide effort. Its bold "billion dollar" aspiration is not as audacious as it might first appear. Their preliminary research indicates that college endowments can easily generate the sufficient start-up capital. Indeed, the Founding Circle of BDGC members includes 32 colleges and universities that have already set up 65 million dollars of GRFs.

The Billion Dollar Green Challenge publishes several guides that offer instruction and support. These include the capstone report, *Greening the Bottom Line: The Trend toward Green Revolving Funds on Campus* and the very helpful case studies of exemplary GRFs. *A Green Revolving Fund Investment Primer* provides additional guidance, as does *The GRK Implementation Guide: Step-by-Step Strategies for Financing Energy Efficiency*.

The most useful service is the Web-based Green Revolving Investment Tracking System (GRITS), which provides real-time comparative data on the performance of GRFs at participating institutions. This best-practices approach epitomizes the collaborative power of GRFs. An important obstacle for many sustainability initiatives is the lack of financial capital for new projects. The GRF program with its technical support is designed to provide the shared experience, expertise, and know-how to overcome any financial obstacles.[29]

GRFs are an effective way to build collaborative partnerships involving colleges, businesses, and communities. As campuses usher in a new era of "green" research, innovations, and investments, they can catalyze "green" community development. Towns and cities have similar infrastructure challenges. Businesses have ready-made markets for "green" products and processes. Together they can create an economy of scale for investments, workforce development, and sustainable entrepreneurship. Further, such partnerships allow for dynamic, profit-based capital investments. It will be of great benefit to campuses nationwide if they can share their experiences, compare financial tools, better understand returns on investment, and optimize their innovative technological solutions.

The revolving loan fund might stimulate investments for broader community partnerships. With slight modifications, and then linked to regional capital alliances, it can provide a campus (and a community) with diverse revenue streams—the foundation of resilient sustainable prosperity. Similarly, inter-campus collaborations can accrue even more investment potential. Revolving loan funds can also seed profit-oriented businesses, linked to campus/community initiatives.

There are major curricular implications as well. Students, staffers, and faculty members who administer and implement GRFs get an instant lesson in business planning, infrastructure development, and community engagement. It is another instructive way to get business schools directly involved in the financial challenges of their campuses and communities. The Billion Dollar Green Challenge is a stimulus for campus sustainability initiatives. It is a useful investment strategy for building sustainability-based revenue streams.

The Campus as Sustainability Incubator

One of the biggest challenges for educational institutions is to find ways of diversifying their revenue streams. As I described earlier in the case of Unity College, many American colleges and universities rely on tuition as the most important source of revenue. Colleges and universities also rely on endowment revenue, but that can't be counted on as much in a volatile economy. Historically, state colleges and universities could rely on the state legislature to grant ample budget requests, but in times of budgetary cutbacks those allotments are shrinking dramatically. The broader philanthropic revenue stream is increasingly competitive, as are various grants and awards.

A sustainability ethos will succeed only if it can support its initiatives by adding to the campus revenue stream. The revolving loan funds are an ingenious mechanism for redirecting existing revenue and maximizing the returns on it. They can also be used as investment leverage, or seed capital, for stimulating campus-related business opportunities.

The term "business incubator" is used to describe a process that supports nascent entrepreneurial enterprises. Several campuses have initiated sustainable business incubators[30]—programs that support the development of new sustainable businesses. College campuses are incubators, too, providing an educational environment that allows people, ideas, and projects the nourishment to grow and flourish. Extending this idea sequence further, consider the campus as a sustainability incubator, an environment that allows diverse forms of campus eco-capital to emerge as long-term wealth assets.

Combine the sustainability incubator with the three elements of infrastructure—energy, food, and materials. In each case, the campus can engage in capital investments that will utilize the varieties of campus capital (financial, natural, social, intellectual) to build long-term campus wealth. Energy investments in retrofitting and in alternative energy sources will have financial benefits, stimulate research, and bring prestige and interest to the campus while also enhancing teaching and learning. Food-growing programs can also save money, build community partnerships (social capital), employ students, and become the basis for innovative curriculum and research. Innovative use of materials reduces pollution, improves the quality of life, saves money, and is another source of research and teaching. A campus is an ideal place to invest in these projects and to demonstrate how they enhance value by utilizing diverse forms of capital. Successful projects in each of these areas support sustainability curriculum in many fields, including business, energy, agriculture, materials science, and ecology. They contribute to economic renewal, workforce preparation, community partnerships, and innovative research—the essence of sustainable investment.

The Campus as Regional Sustainability Hub

Through innovative investment partnerships, a campus has a magnificent opportunity to incorporate its varieties of capital—financial, intellectual, social, and natural. By so doing, it has the potential to become a regional sustainability hub, or convener. By virtue of its financial capital, the campus invests in sustainable community initiatives. It uses its intellectual capital to invest in dynamic research programs for sustainability science and innovative technology. Through its social capital, it builds resilient community partnerships and integrates the region's schools and learning centers. Natural capital allows the campus to become a regional energy and food provider, as well as an ecosystem service provider.

All these forms of capital incubation allow the campus and the community to cultivate diverse revenue streams. The financial capital is organized to eliminate debt, to employ students (helping with tuition costs) and members of the community, and to build revenue through successful sustainable business ventures. The intellectual capital builds revenue through think tanks, consultancies, consortia, and conferences. Social capital facilitates revenue growth through social marketing, improves the quality of life, and broadens the understanding and awareness of the sustainability ethos. Natural capital allows the campus to maximize its ecologically sound use of natural resources, providing local energy, food, and materials, while maintaining biodiversity and landscape integrity.

These projects, viewed as a comprehensive, long-term, strategic sustainability portfolio of possibilities, promote the campus as a dynamic educational enterprise, closely linked to the community, partnering with businesses, and ultimately enhancing affordability, access, and significance.

Although no campus has yet to create such a comprehensive orientation, many are organizing projects conducive to dynamic campus and community investments. The Tompkins County Climate Protection Initiative) is a "multisector collaboration seeking to leverage climate action commitments made by Cornell University, Ithaca College, Tompkins Cortland Community College, Tompkins County, the City of Ithaca, and the Town of Ithaca to mobilize a countywide energy-efficiency effort and accelerate the transition to a clean energy economy."[31] Arizona State University's Global Institute of Sustainability "focuses on research that is useful to the greater community from Phoenix to around the world.

The Institute also connects researchers with practitioners from business, industry, municipalities, and government to collaborate on planning for sustainability challenges of urban growth, environmental protection, resource management, and social and economic development."[32] A survey of college and university sustainability programs will reveal dozens of these efforts that aspire to position the host institution as a regional sustainability hub.

A Thriving Sustainability Marketplace

The college campus serves as a dynamic economic multiplier. On a large campus, many financial decisions, including routine procurement contracts, food and hospitality services, and the provision of energy, can make an enormous difference for the business that gains the contract. When an institution uses sustainability criteria to make its decision, and then awards the contract to firms that emphasize those criteria, it is supporting the emerging "green" economy.

Siemens, Honeywell, Aramark, and many other companies have divisions that emphasize "green" products and systems suitable for larger colleges and universities. For example, energy service contracts can be very lucrative. On a smaller scale, when a campus awards a contract to a small business, it can provide an important boost to the prospects of that business. At Unity College, we took great pride in having our own food service and in making food-purchasing decisions that supported the local economy. We couldn't always do this, and there were times when a national distributor provided much better service. However, we knew that our contracts with local "green" businesses could make a crucial difference for regional sustainability efforts. Our intention was to use our purchasing power (as small and resource strapped as we were) to promote secure and stable sustainability markets.

Some campuses have held sustainable business fairs at which "green" businesses, organizations, and non-profits have demonstrated their wares. Such fairs present wonderful opportunities for students to be considered for jobs and/or internships. Campus programs can also have booths at these fairs. In urban centers, this is an effective way for students to get a sense of the opportunities available in their metropolitan area, and it allows the participating organizations a chance to sell their wares or recruit talent. In rural Maine, Unity College would always attend the Maine Organic Growers and Farmers Association's Common Ground Fair, one of the largest agriculture-and-sustainability-oriented gatherings in the United States and an outstanding example of a thriving sustainability market.

Campus leaders can work with the faculty to invite CEOs of regional and national businesses to come to the campus for discussions about workforce preparation. The businesses can describe the kinds of skills they require, the kind of work their employees are most likely to engage in, and the business environment they need to grow a successful sustainable business. In exchange for courses and/or programs that will facilitate such preparation, the businesses can provide apprenticeships, scholarships, and other means of support.

Colleges and universities can support a "green" economy by considering their campus environment as a thriving sustainability marketplace. The campus is a foundation for sustainability innovation, a locus of community opportunity, and a center for workforce preparation. Innovation occurs through research, experimentation, and implementation, using campus operations as a means of generating sustainable solutions. When the campus is seen as an opportunity generator, it demonstrates how and why its research and curriculum is pertinent. Career preparedness is enhanced when the curriculum is concretely connected to the emerging job market. The sustainability ethos will maximize its influence and its effect when campuses include these important economic considerations.

6

Wellness

Wellness as Human Flourishing

To stay in good physical shape, and to rejuvenate my energy, I always take a mid-day exercise break. When I was at Unity College, I enjoyed taking a lunchtime bicycle ride. For the first few weeks I went solo. Several of my senior staffers noticed and they asked if they could join me. Before long, we had a cohort of riders, including students, staffers, and faculty members. I'd send an e-mail message to the extended cohort each morning announcing the time and place for the daily ride. This became a wonderful routine. I had some of my best meetings while bicycle riding through the rural Maine countryside.

The entire campus (and the town) soon took notice. Before long, many of the staffers started taking exercise breaks during the day. We considered how the college could support a comprehensive approach to campus fitness. The vice president of academic affairs, the dean of student life, the athletic director, and I (all members of the lunchtime bicycle team) hatched plans for supporting "wellness" initiatives on campus. We introduced nutritional awareness, mindfulness meditation, frisbee golf, seasonal vaccinations, and wilderness outings. I am convinced that this boosted campus morale, improved community health, promoted a sense of good will, and set a public example. I wanted to promote not only the virtues of physical fitness, but also the necessity of living a balanced, healthy, and joyful life.

Sustainability and wellness are inextricably linked. When you strip away all the layers of the nine elements and you seek the origins, motivations, and intentions of the sustainability ethos, what is the essence of its deeper meaning? I believe it is the aspiration to live a fulfilling life. What words shall we use to describe this aspiration: wellness? well-being? human flourishing? happiness? exuberance? celebration? And what virtues are the foundations of a good life? Which virtues are intrinsic to sustainability, and how are they linked to campus life? These are complex questions of meaning and purpose. People are attracted to the sustainability ethos for countless reasons, and it would take many late evenings of good conversation to better understand various interpretations of human flourishing. Still, the issue must be addressed. Aspirants and critics alike assert that sustainability encompasses values. What better place is there to discuss values than a college campus? The sustainability ethos requires a philosophy of education. It implies that we think carefully about the meaning of a good life. Late evening conversations about such topics are a very good thing.

These conversations range far and wide, through spiritual and scientific territory, through the realms of ethics, art, history, and psychology, spinning diverse narratives, raising intricate questions, probing all manner of inquiry and experience. Yet there is a core belief that distinguishes the sustainability ethos. This is the assertion of a correspondence between human flourishing and ecological resilience, that the future of humanity and the biosphere share a common destiny, and that personal well-being is connected to the fate of the planet. This challenging prospect must somehow be reduced, applied, and internalized within the boundaries of common everyday experience. To do so, we strive to attach human flourishing to local ecology, and we make this tangible by illuminating the ecological impacts of daily life practice. This informs how we think, learn, and live. I believe it is the essence of the sustainability ethos.

Wellness may be described as the many dimensions of human flourishing,[1] situated in an ecological context. The sustainability ethos considers wellness by simultaneously linking and assessing the common health of the individual, community, and the ecosystem. Its overriding concern is that ecosystem processes are either poorly understood, ignored, or perceived as an auxiliary sink, an unlimited depository for human waste. Its premise is that ecosystem processes must be deeply understood, internalized, and perceived as intrinsic to the human quality of life, with respect for the capacity and dynamics of the biosphere. Hence human flourishing and ecosystem processes are intrinsically interconnected, representing a reciprocal, co-evolutionary relationship.[2]

By what criteria shall we assess wellness? And how might wellness be connected to campus life, especially as sifted through a sustainability filter? We can approach these questions by further considering the conditions that allow for human flourishing. Campus wellness is a subjective measure of those conditions, as reflected in personal health and fitness, community purpose and vitality, and ecological resilience.

The sustainability ethos should prompt lively conversations about the meaning of fitness, the importance of wellness, and the various dimensions of human flourishing. I will initiate some elements of the conversation in this chapter, emphasizing its relevance for sustainable campus life. In this chapter, I'll consider what it means to have a healthy campus, the signs of community vitality, and how we might respond to some common sources of institutional stress. I'll suggest that sustainability initiatives can contribute to community vitality and why the resilience concept is a useful strategic approach for connecting personal and organizational wellness to broader ecosystem challenges. I'll explain why a place-based orientation grounds a campus in its local ecology and calls attention to the various ways a campus serves as home. A healthy campus can use the techniques of restorative environmental design to connect place, sustainability, and community health. Many universities are using service learning as a way to promote community vitality by linking the campus to the region through their combined good works. This is a fitting venue for sustainability initiatives. In the final sections of the chapter, I'll describe the relationship between sustainability, character, virtue, and life practice.

Fitness and the Campus

Here are some common characteristics that often contribute to feeling well: excellent physical and mental energy, minimal bodily aches and pains, a reasonably relaxed state of mind, a sense of enthusiasm about our daily activities, a sense of purpose regarding our work and its effect, a work environment that allows both autonomy and collaboration, and knowing one has supportive friends and family members and that they are thriving.

But many things can intervene: sub-optimal energy, annoying aches and pains, feeling decidedly unrelaxed, a lack of enthusiasm, no real sense of purpose about one's work, loneliness, an oppressive work environment, worries about friends and family members. Then there are all the unexpected interventions that may occur. The extraordinary psychological challenge of the well-adjusted individual is to balance these ups and downs, place them in perspective, and still maintain optimism, confidence, and gratitude in the face of uncertainty and the waves of good and bad circumstances.

These challenges are discussed in the deepest spiritual texts and in popular self-help books. Yet wellness is often neglected in discussions of sustainability. I'm not sure why this is so. Perhaps the daunting challenges of our planetary predicament render discussions of wellness inappropriate. Or maybe it is hard to overcome the implicit psychological tensions of modern campus life—the idea that you are most productive when you are at your busiest and most stressed. In my view, wellness (human flourishing) is the essence of the sustainability ethos, and if we don't practice living well we can easily lose our bearings and we ultimately diminish our effectiveness and health.

Fitness is an important characteristic of wellness. Recently exercise mavens have usurped the word "fitness," and it now conveys being in the best possible physical shape. Darwin, however, used fitness to describe the ability to survive and reproduce as criteria for natural selection. It is helpful to remind ourselves that fitness has a biological origin, especially as we link individual and ecological well-being. I am using "fitness" in the classical sense, as a description of an individual's overall well-being. Fitness is an aspiration as much as a condition.

Assessing fitness requires a context. What is fitness for, and in what setting are we considering it? In regard to campus life, the well-being of all community members matters—students, staffers, members of the faculty, senior leadership, visitors, and alumni. How does a campus provide the services and venues to allow for the potential of physiological and psychological fitness?

For a campus to support physiological fitness, several qualities stand out—access to affordable health care, nutritious cuisine, spaces for physical exercise, and access to recreational ecosystem services. These benefits or services provide a synergy of effects, and together they maximize both individual and campus well-being. On a college campus it is equally important to educate the community about their meaning and importance. To what extent are discussions of these benefits embedded in the campus curriculum? How is the curriculum tied to everyday life practices? I will address these questions in chapter 7. Here I will simply note that any college campus that provides these benefits is likely to be a healthier place, and that it will more successfully recruit and retain students, staff members, and faculty members.

A campus can better ensure psychological fitness by maintaining an atmosphere that values meaningful work and study, a sense of community belonging, multiple opportunities for personal growth, outlets for creativity and personal expression, places for leisure and recreation, and accessible counseling services. A campus has a responsibility to provide the conditions for maximizing psychological well-being. It can't necessarily deal with the various idiosyncratic situations that may emerge, although it should be equipped to refer people to qualified professionals as appropriate. One of the most profound challenges for college campuses is dealing with the complex range of psychological issues that negatively affect learning and community life. A campus can help address these issues by building a supportive and nurturing community.

Community Vitality

Any astute observer, upon entering a campus for the first time, will gain enduring first impressions based on the spirit, morale, warmth, and vitality of the environment. How are people welcomed? Can they find their destinations easily? Do the students, staff, and faculty members seem engaged, enthusiastic, and inviting? How does it "feel" to be a visitor to the campus? Is the campus welcoming, interesting, and dynamic? First impressions, lasting though they may be, are not always correct, and they may not be consistent. That's why it is helpful to have systematic quality-of-life studies that enable a campus community to deepen its understanding of morale,[3] spirit, and satisfaction.

College and university campuses are complex communities, ranging from very small, almost commune-like environments to sprawling cityscapes with tall buildings, museums, stadiums, concert halls, theaters, laboratories, and residential towers. A campus may consist of multiple communities, with commuters, residents, employees, visitors, and attendants. They are unique, too, in that their student residents are often transient (depending on the length of their program), and the employees may range from part-timers to lifers. The people who live and work there may be primarily local or regional, or they may come from diverse international settings. Indeed, college and university campuses represent an intricate breadth of community matrices. This is why they are so interesting. Why are some campus communities more vital than others?

Oregon Explorer (based at Oregon State University), a project designed to promote community engagement, defines community vitality as "the ability of a community to sustain itself into the future as well as provide opportunities for its residents to pursue their own life goals and the ability of residents to experience positive life outcomes. More specifically, we suggest that a vital community has community capacity (the ability to plan, make decisions, and act together) and realizes positive social, economic, and environmental outcomes."[4] Most campuses and universities probably share a similar aspiration and promote themselves accordingly, and acknowledge the close relationship between personal well-being and community well-being. Further, they encourage a common sense of purpose—the importance of education. Although their approaches to education may differ, and they may have other institutional purposes, I don't think I have ever encountered a campus that didn't include various measures of community vitality and personal well-being as intrinsic to its educational mission. Indeed, these measures rank highly on the standards of most higher-education accrediting bodies. Yet not all campuses are equally vital, nor are all campuses healthy places to work and study.

What are some of the factors that contribute to community vitality? When community members participate in a compelling vision and purpose that lends meaning, voice, and identity to their work, they are more likely to have positive feelings about the campus. When people work collaboratively in pursuit of common goals, they are more likely to foster support, encouragement and confidence, leading to a sense of enthusiasm, dynamism, and spirit.

These prerequisites for community vitality seem obvious, but the implementation is challenging. Community building can be subverted by the myopic pursuit of individual goals, unilateral decision-making processes, unexpected budgetary constraints, the inability to find common purpose, and other breakdowns of trust, communication, and conviviality.

Sources of Stress

College and university campuses strive to provide healthy learning environments, yet they are often stressful places. Whether a campus primarily serves a traditional undergraduate population (18–25-year-olds) or adult learners (25 and over), it is a nexus for anticipation, aspiration, challenge, and change. College freshman are typically experiencing new forms of independence. They are exposed to many new ways of thinking, various cultural and political perspectives, and numerous choices, temptations, and possibilities. Developmentally, these choices correspond to an important stage of identity formation.[5] The college environment places additional pressures on students, from academic expectations to peer-group approval. Economic pressures may exacerbate these tensions. The great majority of undergraduates take on a cumbersome, even overwhelming debt burden, often necessitating rigorous work schedules and placing their families under considerable stress. It is not surprising that student life staff and administrators must provide effective counseling psychology services. This is becoming a huge burden (and expense) for many campus communities.

Adult learners face equally challenging circumstances. The economic considerations include having to manage part-time or full-time work, families, a similar potential debt burden, and all the doubts that accompany their decision to return to school, especially the uncertainty of whether their time and investment will result in a meaningful career with good employment prospects. Despite these uncertainties, they choose to pursue their education because of the prospects for career advancement, including educational growth, and personal fulfillment. This is the great promise that motivates their participation. Advertisements for colleges and universities typically promote the promise of accomplishment and success and feature images of successful people engaged in meaningful activities.

Personal stress issues contribute to organizational stress. The campus has an obligation to deliver on its educational promises. This is difficult enough during periods of relative economic affluence. When there is economic uncertainty, additional burdens are placed on colleges and universities, most critically in terms of budgetary cutbacks, financial restrictions, and public accountability. As I write this, extraordinary budget cuts are facing California's state university system, threatening the closing of programs, larger class sizes, layoffs, and the termination of many important educational services. This is a nationwide challenge, placing additional organizational stress on campuses as members of the staff and the faculty face workload challenges, job-security challenges, and quality-of-working-life challenges. At their gatherings, college and university leaders often discuss how to manage the organizational stress that accompanies this economic reality. They are especially concerned with the psychological implications of these complex financial burdens, both in terms of organizational and institutional stability and in terms of the effect on all members of the campus community.

As college and universities strive to provide a nourishing, supportive, sustainable approach to campus living, they are faced with challenging resource issues. It is important to emphasize that these observations are generalized, and that campus cultures face both common and specific challenges. Many campuses are remarkably sheltered and conducive to identity formation, stress reduction, and unprecedented educational opportunity, and they successfully maintain those standards. American colleges and universities have vastly different resource capacities. Yet many campuses are striving to develop viable financial models, and may be unable to provide protective hedges against the prevailing financial challenges.

Sustaining a Healthy Campus

How is it possible to sustain a healthy campus in view of these multiple individual and organizational stresses? How can a campus promote organizational resilience in the midst of changing economic circumstances and dynamic organizational change? What is the obligation of the campus in providing the conditions for human flourishing—a nurturing, vital, meaningful, and purposeful community for pursuing educational opportunity? How might members of the campus community work together to diminish stress?

An emphasis on physiological health and psychological well-being provides a learning opportunity for an entire campus. As such, a "wellness" strategy may be embedded in all aspects of campus planning, from curriculum to quality of working life, becoming a priority for the board of trustees, the senior leadership, the staff, and the faculty. Consider the three broad structural clusters of the nine elements—infrastructure, community, and learning.

The physical infrastructure of a campus is the interface between the built environment and the ecosystem. Many of the most effective sustainability initiatives emphasize the convergence between ecosystem health and daily life practice. For example, one can have an enormous effect on campus wellness by promoting awareness of whole food and nutrition, and linking such awareness to how food is grown, distributed, prepared, and served, and integrating such concepts into all aspects of student life. There are similar convergences with energy use and materials use, water, the uses of green spaces, landscaping, lighting, and transportation. A healthy campus promotes sustainable practices with all these functions.

Community vitality is optimized when senior leadership builds partnerships among all campus constituencies, emphasizing the quality of working life and the importance of psychological well-being. Austerity creates difficult challenges, but it also opens opportunities for partnerships, transparency, and collective effort. Sustainability initiatives emphasizing wellness can unify a campus. Various departments, especially the Student Life, Public Relations, and College Advancement departments, can work together to promote the campus as a center for community wellness. Many campuses sponsor community partnerships that cultivate healthy life practices—for example, bicycle commuting, farmers' markets, athletic activities, environmental field trips, and community meals. Senior staff members can take the lead in organizing, promoting, and participating in these activities. Wellness is then built into the mission, reputation, and stature of the campus.

Wellness initiatives may be integrated in both the daily life practice of the campus and the curricular agenda. From freshman orientation to graduate seminars in professional schools, the study of wellness and its relationship to sustainability provides ideal material for educational programs. Interesting areas for research and teaching include the neurophysiology of emotional resilience, ecological resilience, human adaptation, nutrition and academic performance, social capital, and quality of life.

Most campus environments are responding to the stresses of global and local environmental change, a precarious global economy, and the accompanying social, cultural and psychological challenges. Perhaps the ultimate measure of campus wellness and the prospects for human flourishing will be a campus's resilience in three domains. Emotional resilience describes an individual's capacity to manage change. Organizational resilience describes how well governance adapts to and leads change processes. Ecological resilience describes how the campus manages its ecosystem services and responds to environmental change.

Ecological Resilience and Community Vitality

The emerging concept of ecological resilience reflects a growing awareness of and concern about rapid environmental change. Ecologists and climate scientists require concepts and criteria for understanding how to study, mitigate, respond, and adapt to these changes, especially those with potentially catastrophic impacts. Ecological resilience is "the capacity of an ecosystem to respond to a perturbation or disturbance by resisting damage and recovering quickly."[6] It is an important component of ecosystem health, community vitality, and campus fitness.

The ecologists Brian Walker and C. S. Holling describe four aspects of resilience: latitude (how much a system can change before losing its ability to recover), resistance (how easy or difficult it is to change a system), precariousness (how close a system is to its threshold), and panarchy (how much a system is influenced by other systems). These highly abstract concepts require tangible explanation and expansion, most effectively accomplished through ecological and organismic case studies. Such elaboration is beyond the scope of this discussion. Nevertheless, it is important to introduce these concepts, as they provide a foundation for sustainability practices via scientific ecology.

Interestingly, the psychologist Richard Davidson, in a book titled *The Emotional Life of the Brain*, describes resilience as one of six criteria in what amounts to a neuropsychological emotional health system or, stated more simply, one's "emotional style." Essentially, resilience in psychology "refers to the idea of an individual's tendency to cope with stress and adversity."[7] Davidson describes the neurophysiological evidence suggesting that resilience is fundamental to human emotional health.

For our purposes, it is metaphorically suggestive to consider the various ways we can use resilience as a descriptor for health, particularly as fundamental to sustainability and wellness. A campus community ostensibly seeks to maximize human flourishing under conditions of rapid environmental change. "Resilience" describes their ability to flexibly adapt to the changing circumstances of their lives.[8] By connecting ecosystem resilience to emotional resilience, we attest to the correspondence between people and the planet. We suppose that the qualities of resilience are a prerequisite for personal, community, and ecosystem health. Community vitality is the substrate of human flourishing, the terrain from which collective aspirations arise, and the home place where personal and campus fitness encounters ecological resilience via the habits and routines of daily life.

"We All Live Here"

A campus is rooted in the cultural landscape of the ecological region. Human flourishing is more likely to be sustained when there is a smooth interface and respect between people and place. An interesting challenge for any campus community is how to create a sense of place and a temporary home for all of the people who are just passing through—students who may reside there for short periods of time, staffers and faculty members who work there and live elsewhere, board members who may fly in from distant regions. A highly functioning campus will feel like "home" to its constituents. Why else is there a "homecoming week"?

A thriving campus is a diverse, even cosmopolitan, community center where many different people feel at home. Consider the various ecological and cultural forms of campus diversity—intergenerational, intercultural, international, and interspecies. Perhaps the ultimate criterion for assessing a campus's wellness and vitality is the extent to which such heterogeneity can thrive there. Similarly, campus heterogeneity may also reflect the depth, pertinence, and vitality of the educational programs.

An intergenerational campus provides a home for people of all ages. welcoming students with diverse educational interests and challenges, who may be at very different stages in their careers. This can be a wonderful educational opportunity, supporting two-way mentoring and access to a variety of life experiences. Classroom discussions are often more interesting when different generations of students interact. Perhaps more intergenerational diversity can also provide some relief for campus psychological services, as individuals with interesting life experiences offer helpful counsel to one another.

An intercultural campus brings people of various ethnicities, belief systems, and social backgrounds together. Although this creates communication challenges, it offers an unparalleled way for students to experience what they are most likely to encounter in their working lives. It is essential that they learn to understand, tolerate, and respect difference, and in so doing, they open themselves to new avenues of cultural learning. Minority students, in particular, face acculturation issues on many campuses. When these issues are handled with sensitivity, compassion, and support, the student is more likely to feel "at home" on campus, and perhaps more likely to graduate. These acculturation issues reflect another dimension of human flourishing. Can the whole campus provide support for acculturation and thereby maximize the learning experience for diverse populations?

Many campuses are exploring new international partnerships as a means of generating additional and much-needed revenue. Hundreds of American colleges and universities (including many rural institutions) now have substantial populations of international students. As campuses expand their reach internationally, they are more likely to generate intercultural learning opportunities. An international campus may consider new educational programs that meet the needs of a global student body. Students who are very far from home require unique forms of counseling and psychological support. How might a student from a Vietnamese city adjust to campus life in a rural Virginia community? How does a student from rural China adjust to life in downtown New Haven? A thriving international campus is a home for people from many different countries.

An interspecies campus promotes biological diversity, maintaining existing habitats for local flora and fauna, restoring such habitats as appropriate, and cultivating an ecological landscape for food growing, wildflowers, and bird and butterfly migration.[9] The sustainability ethos encompasses the integration of the built environment and the ecosystem, suggesting that biodiversity is a primary indicator of ecosystem health and resilience. Hence a campus should take pride in the number of species that share the ecological community of its place. The campus community, broadened to include the ecological region, encompasses more than just humans. An interspecies campus, with healthy biodiversity, reminds the community of its place in the ecosystem.

A thriving place integrates cultural diversity, biodiversity, and a sustainable community.[10] There is growing scholarly evidence from contemporary case studies and traditional ecological knowledge suggesting that biodiversity is more likely to be sustained in culturally diverse regions. The UNESCO report *Cultural Diversity and Biodiversity for Sustainable Development* is based on the premise that "respect for biodiversity implies respect for human diversity." It continues: "Cultural diversity—as a source for innovation, creativity, and exchange—is the key to a mutually enriching future for humankind. Cultural diversity guarantees sustainability because it binds universal developmental goals to plausible and specific moral visions. Biological diversity provides an enabling environment for it."[11]

By enhancing these dimensions of diversity, the campus environment becomes an exemplar of community vitality, cosmopolitan heterogeneity, and a manifestation of local place in a global context. It demonstrates a willingness to incorporate an ecological and cultural vision of human flourishing. By acknowledging its significance as a temporary home for various populations, it embodies the profound challenges of living together in a complex world.[12] What better model is there for curricular relevance, practical experience, and a healthy campus community?

I'll continue this conversation in chapter 7. Here, I'd like to reiterate the importance of home place. This understanding is enhanced when it is coordinated with how we conceive and use buildings. One of the most exciting new approaches to campus wellness is the incorporation of restorative environmental design into the built environment.

Restorative Environmental Design

Principles of ecological design incorporate community health with ecosystem awareness. In the last ten years, there have been scores of dramatic, imaginative, and practical ideas regarding campus architecture, master planning, and landscape design, with emphasis on low-impact energy and water use, conservation, and recycled materials. The theoretical literature is vast. The practical applications are resplendent. A survey of exemplary "green" campus buildings and landscapes yields a stunning portfolio. These efforts clearly are oriented not only to provide attractive and functional working and living spaces, but also to promote personal, community, and ecological health.

Despite these fine efforts, we have only scratched the surface of what is possible. Although some campuses cite budgetary constraints as a limiting factor, there are countless low-cost approaches to these kinds of sustainability initiatives. The most important challenge is conceptual—how to build principles of ecological design into all aspects of campus planning. It is helpful to reiterate that there is a considerable body of research that demonstrates the direct correspondence between ecological design and human flourishing. This awareness should inform all campus decision makers.

Stephen Kellert synthesizes some of the best work in this regard in his book *Building for Life*. Kellert suggests the rubric "restorative environmental design" because it "incorporates the complementary goals of minimizing harm and damage to natural systems and human health as well as enriching the human, mind, body and spirit by fostering positive experiences of nature in the built environment."[13]

In Kellert's scheme, "(1) Low environmental impact design sustains various ecosystem services on which human existence relies, (2) organic design fosters various benefits people derive from their tendency to value nature (biophilia), and (3) vernacular design enables a satisfying connection to the places where people live, also a necessary condition of human well-being."[14]

Kellert describes organic and vernacular design as components of "biophilic design," grounding the scheme in E. O. Wilson's concept of biophilia—the hypothesis that there is an instinctive bond between human beings and the natural world. He describes eight elements of biophilic design: prospect (ability to see into distance), refuge (sense of enclosure or shelter), water (indoors or inside views), biodiversity, sensory variability, biomimcry, sense of playfulness, and enticement.[15] He concludes with a list of design principles that should inform a "new way of thinking about building and landscape design."[16] These include incorporating biophilic values in design processes, explicit linking of material and energy flows to spatial and temporal biogeographical scale, an emphasis on interdisciplinary perspectives in master planning processes, measuring performance standards, and building campus awareness and appreciation of restorative environmental design.

What I find most appealing about Kellert's approach is the integration of all aspects of campus life. Promoting public awareness of these design principles, which he emphasizes, is superb advice for any campus that undertakes sustainability initiatives. Restorative environmental design is a conceptual foundation for not only master planning and building construction, but for all aspects of the campus community—from the basic practices of daily living on campus to the institution's curricular objectives. In Kellert's view, biophilic design reiterates a primary evolutionary relationship: humans require access to natural settings in order to maximize their health. His strategic orientation is clear. Most people spend the majority of their time in built environments and most people live in cities. How do we simultaneously maximize ecosystem health and provide people and communities with opportunities to experience the natural world? Kellert suggests we do so by integrating the ecological features of the landscape into all aspects of our buildings and grounds.

This approach can be neatly synchronized with the concept "we all live here." The interspecies matrix of the campus provides the template for how we live in a place. The diverse cultural inhabitants bring their unique perspectives to how this cohabitation might occur. The multi-generational members of the community share their memories and aspirations about how people lived in the place at different moments of its history (Kellert's vernacular design). I can't think of a better way to bring these groups together in their campus home than to gather their stories and experiences about how to live well in their shared place.

The Necessity of Service

Service learning plays an increasingly important role on many campuses, serving various curricular purposes. Hundreds of colleges and universities emphasize service learning in their promotional materials and support its delivery as crucial to their academic programs. The assumption is that service promotes civic participation, provides meaning and purpose for scholarship, emphasizes hands-on learning, and bridges the gap between school and community. Inherent in these assumptions is that service enhances virtue and builds social capital. Further, it is assumed that community service is a strong indicator of community wellness.

This emphasis is so prevalent that it receives bipartisan political support in Washington. The President's Higher Education Community Service Honor Roll was launched in 2006 for the purpose of recognizing campuses that successfully place students on "lifelong paths of civic engagement." The program's website (as of 2012) asserts that there are 2.5 million students involved in community service, reflecting 105 million community service hours, approximating a value of $2.2 billion![17]

Regardless of how we measure the relative worth of these activities, it is safe to assert that community service is perceived as a highly valued attribute of campus life. The most recent findings of the National Survey of Student Engagement support this emphasis, especially for first-year college students: "For new students, service learning creates meaningful connections with the community and deeper interactions with faculty members and peers while enhancing their sense of civic responsibility."[18] Further, according to NSSE, participants in service learning are more likely to work effectively with others, understand themselves, understand people of other racial and ethnic backgrounds, solve complex real-world problems, develop a personal code of values and ethics, and contribute to the welfare of their community.

At Unity College, we placed great emphasis on service learning as intrinsic to improve retention and maximize engagement. Our assumption was that retention and engagement were reliable indicators of a campus's vitality and thus essential to our conception of wellness. Although an increased reliance on service learning was one of many wellness initiatives, there was a direct correlation between improved retention and comprehensive opportunities for service learning. A survey of colleges and universities in northern New England sponsored by the Maine Campus Compact concluded that "students who participated in service-learning courses . . . reported higher community engagement, academic engagement, interpersonal engagement, academic challenge, and likelihood to remain at the university (retention) than students in courses that did not include service learning." In addition, "a mediation model showed that academic challenge and academic engagement were the elements of service-learning courses that most influenced students' decision to stay at the university (retention)."[19]

Service learning integrates curriculum, campus life, and community partnerships. It blends seamlessly with campus sustainability initiatives. At Unity College we developed a robust portfolio of such opportunities, providing students, staffers, and faculty members with many options—growing organic food in partnership with regional hunger organizations, supporting energy retrofits for low-income housing, participating in research on arsenic in the local water supply, organizing Maine-based sustainability service field trips for spring break, organizing regional gatherings on the meaning of place, and partnering with local farmers to establish community markets. Service learning is the ideal venue for integrating sustainability theory and practice, building partnerships between the campus and community, uplifting morale and the quality of working life, and thus contributing to a healthy campus. There can (and should) be a service component linked to each of the nine elements of a sustainable campus.

Networks That Matter

Service learning highlights the importance of face-to-face relationships while strengthening social connectedness. It establishes robust and enduring partnership networks. These networks may address social service needs, environmental concerns, planning challenges, or any issue that is important to the campus and the community. These are networks that matter.

Many campuses are trying to internalize and incorporate the rapid proliferation and popularity of electronic social communication networks. Predictably, administrators are striving to understand how these networks affect student success and engagement, including retention, study habits, and preparation. These are complicated discussions, eliciting multiple perspectives regarding the relative benefits and detriments of these developments. Do Facebook, Twitter, and other such networks bring people together, further isolate them, allow for new approaches to learning, or serve as insidious distractors? A lot of evidence supports the use of such networks; a lot of evidence criticizes it. Unless one is willing to ban their use (not an entirely bad idea for short periods of time), they are simply a fact of campus life.

Electronic social networks are simultaneously significant and mundane. They are significant in that they change how we perceive the nature of community, and mundane in that they can serve to endlessly recreate gossip loops and trivia. At their very best they provide spontaneous access to great conversations; at their worst they create distortion, gossip, misattributions, misunderstanding, and even pathology. I'm sure every campus experiences the full gamut of these possibilities. What is unique about electronic communications is how quickly a flurry of messages can spiral into unintended consequences. Campuses attempt to deal with these challenges by developing technical and social protocols for the use of networks. However internet technology opens multiple communication channels. These networks really can't be controlled.

However, as criteria for wellness, a campus can emphasize networks that provide meaningful academic and social engagement. Participation in those networks will yield their own reward. Which networks build social capital? Does the campus provide or support networks that coordinate service learning, sustainability initiatives, community partnerships, physical and psychological health, career and academic counseling, and access to other relevant and necessary services? Is the community sufficiently aware of these networks? Do they reflect a good balance of electronic communication links and face-to-face venues, a balance of the virtual and visceral? Such networks enhance the stock of social capital and open important communication pathways.

The role of these networks reflects an interesting design imperative, closely related to campus master planning, requiring creative imagination, informed by an understanding of both software innovations and human behavior. An understanding of social networks will be increasingly necessary in any educational or workplace environment.

Sustainability, Character, and Life Practice

Character is higher than intellect. Thinking is the function. Living is the functionary. Ralph Waldo Emerson[20]

In the early 1990s, Thich Nhat Hanh, a Vietnamese Buddhist, conducted a series of meditation workshops oriented to the specific challenges of environmental professionals. I had the good fortune to attend one of them. In my experience during the program and in the twenty years since then, the reverberating expression "you can't take care of the environment if you don't take care of the environmentalist" resided in my awareness. I used it as a way to balance the challenging demands of professional life, to serve as a way to place aspiration and accomplishment in the deeper perspective of a whole life.

Much of the sustainability ethos has its origins in the virtues of simplicity, a vision of a "good life" that has Thoreauvian roots, including "health, freedom, pleasure, friendship, a rich experience, knowledge (of self, nature, and God), reverence, self-culture, and personal achievement."[21] Simplicity reflects an enduring tradition in American history. In a book titled *The Simple Life*, David Shi (a historian who became a college president) reveals the origins and practice of this sensibility. He describes how the simple life was intrinsic to the Progressive movement, including "a cluster of practices and values that have since remained associated with the concept: discriminating consumption, uncluttered living, personal contentment, aesthetic simplicity (including an emphasis on handicrafts), civic virtue, social service, and renewed contact with nature in one form or another."[22]

Sustainability advocates typically support such Thoreauvian values in principle, yet their campus work environments are exceedingly demanding. The sustainability ethos promotes "the good life," but the urgency of the "planetary challenge" and the various stresses of present-day higher education often create pressured and tense work environments. Many campus sustainability professionals I encounter, including staffers, faculty members, managers, and senior leaders, worry about their seemingly unlimited portfolio of urgent and demanding tasks and requests. They are compelled to respond for three main reasons: the perceived importance of the sustainability mission, the motivation to accomplish tangible results, and their desire to uphold standards of personal achievement. This compulsion is stimulated and reinforced by the presumed ubiquity of work, an implicit work ethic, and the assumption that individual and organizational success depends on the exemplary accomplishment of that work. It is relatively rare to find people on college campuses who claim to have achieved a balanced work life. Rather, people complain about, proclaim, or take pride in how busy they are.

We have a profound contradiction here. The sustainability ethos deeply values a "good life" informed by simplicity, communion with nature, and reverence. But the provision of that good life seems to hinder its realization. Of course many people find great satisfaction in sustainability work and find that the work itself is sufficient reward. And how people choose to spend their time and balance their life is an individual matter.[23] Still, my impression, informed by hundreds of conversations with higher-education sustainability professionals, is that for most of these people (regardless of their place and position) there is a fundamental imbalance between the promise of the "good life" and its realization.

What I wish to convey, then, is the inevitable link between sustainability, character, and life practice. Sustainability practitioners are ultimately interested in human flourishing, and serve as the campus conscience for personal health and fitness, community purpose and vitality, and ecological resilience. They are scrutinized because they are espousing ways of thinking, living, and acting. They are expected to model the very behaviors they espouse. As Emerson suggests, how they live and act is as important as what they say.

During my tenure as a college president, I directly confronted this issue. In my role as "supervisor in chief," I had to learn how to create high expectations for the college while espousing a balanced work life. Unless I found the same balance in my own life, I wouldn't be taken seriously in that regard. There was a direct parallel between how I conducted myself publicly and the tone I set for the whole campus. Because I lived on the campus, this was an inescapable reality. Indeed, a modest LEED platinum, zero-carbon presidential residence was constructed to set a public standard for sustainable living. The house functioned simultaneously as private living quarters for the college's president and an educational venue for campus sustainability. Our lives were on display. But the public nature of my life didn't end there.

As a college president, I discovered that people scrutinized everything I did or said. Like many of my peers, I aspired to maximize the educational value of that scrutiny. I won't say that I achieved the balance between high-level professional accomplishment and the sustainable "good life." But I did publicly pronounce my desire to do so, and I attempted transparency in my successes and failures accordingly. I also emphasized the importance of a balanced life for those employees who reported directly to me, and I instructed them to do the same in their departments. At a small college, most work-related complaints eventually arrive on the president's desk. The "well-being" of constituents is always on the president's mind. There is no solace for the president in knowing that a president can't please everyone or that some people just find trouble. And the more accessible and transparent the president is, the more likely it is that people will come to him or her with their issues.

In many respects, the daily challenge of maintaining high morale at a college that espoused the sustainability ethos was the most stressful element of the job. I had to balance the psychological demands of the presidency, my expectations for achieving a sustainable campus, and my aspirations to live and lead "a good life." I contend that this balance is crucial for any sustainability practitioner. The specifics of that balance will vary according to the campus culture, the personal style of the practitioner, and the level of leadership intrinsic to a position. However there are some behavioral tenets for implementing that balance in any institution.

(1) Accept that you are a role model. If you espouse sustainability, people will expect you to live according to your ideals. You can't practice an energy-guzzling lifestyle. If you espouse campus wellness, you probably should eat well, pursue physical fitness, and balance work and play. If you can't do so, then how can you promote it for others?

(2) Provide a sense of proportion and scale. It is often difficult for people to distinguish between working hard and working well. People often misappropriate their time. I spend much of my supervisory time working with people to help them align their priorities accordingly. When you are the supervisor, you are more able to do this. The first question I ask "direct reports" is to tell me how they spend their time, what rationale they use for making their time management decisions, and whether they feel that their work is important. You can't have a balanced working life unless you can figure out how to manage your time.

(3) Emphasize clarity and accountability. Any campus with high aspirations must create a challenging and demanding work environment. How can campus wellness coexist with such aspiration? The key to this balance is requiring clear accountability and expectations. People must know what they can and should expect from one another. The most egregious miscommunications often can be traced to a misunderstanding of who is accountable and what is expected of them. When there is lack of clarity, the stress level in an organization becomes inordinately high. Then you have to spend far too much time (see point 2 above) trying to figure out who was supposed to do what or what people meant when they said something.

(4) Emphasize politeness and respect. This is a simple way to promote a sense of campus well-being. When people treat one another with politeness

and respect, they ensure better communication, they are more likely to speak and listen well, and they will come to every encounter with more confidence and integrity. In contrast, an environment of intimidation, bullying, sarcasm, and condescension promotes anxiety and defensiveness. However, it is crucial that people don't mistake conviviality for a lack of discipline or an unwillingness to set limits. Conflict is inevitable and different perspectives will always emerge. The manner in which conflicts are resolved reflects volumes about campus morale and community vitality.

(5) Create an improvisational flow of creative imagination. I always try to stimulate a creative, improvisational working environment that rewards innovation and imagination. This attitude is absolutely necessary in demanding working environments. It provides an outlet for stress, encourages participation, and demonstrates open-mindedness. Sometimes there are multiple solutions to vexing problems. An improvisational flow doesn't necessarily mitigate a stressful challenge, but it can create more stimulating and rewarding conditions for taking on the challenge. People are most fully engaged in campus life when they are using their imagination to solve challenging problems. An improvisational attitude also suggests there is a willingness to experiment and explore as a way to adapt to changing circumstances.

(6) Remember that purity is the end of possibility. The poet Jim Dodge tells a wonderful story about an experience he once had when a group of students visited the poet Gary Snyder to discuss environmental issues. Snyder served a meal of "road-kill stew" in bowls without silverware. Dodge wondered whether Snyder had "gone Zen pure." Then Snyder went to the kitchen, came back to the dining area, and tossed Hostess Ding-Dongs to the visitors. Dodge suggests that purity is the end of possibility.[24] I have recounted this story on numerous occasions as a reminder that we shouldn't take ourselves too seriously. Our important work requires comedy and lightness.

Reciprocation and Gratitude

Why is Thoreauvian simplicity such an enduring aspiration? For starters, it cuts against the complicated intricacies of contemporary life. In the early nineteenth century, Thoreau conceived a counter to what he considered to be the ubiquitous monotony of daily work life, especially as informed by the routines of commerce. Those routines prevented people from living a full life, mainly by distracting them from direct experience of the natural world. Thoreau's many projects entailed deep immersion in the extraordinary mysteries and intricacies of the immediate landscape, He aspired to shed the shackles of commerce, to roam freely through the fields and forests, and to commit himself to the daily practice of observing nature. Philip Cafaro neatly conveys the essence of this daily practice:

> It is striking how often Thoreau, in discussing the good life, specifies human flourishing and excellence in relation to nature. Some of this is quite basic. The simplest messages in *Walden* are to get outside, use your limbs, and delight in your senses. Run, walk, swim, sweat. Taste the sweetness of the year's first huckleberries and feel the juice dribble down your chin. It feels good to plunge into a pond first thing in the morning and WAKE UP, or to float lazily in a boat along its surface, wafted we know not where by the breeze, gazing up at the clouds. . . . What we need to know in order to live better lives may indeed be very simple.[25]

Nearly 200 years have passed since Thoreau's time. The routines of commerce, the schedules of daily life, the intervening layers of technology, and the expectations of productivity remain considerable. The fields, forests, and ponds are not nearly as accessible. Yet Thoreau's aspirations remain vibrant, and his concept of human flourishing (which also includes the pursuit of knowledge and creative expression) remains relevant. How can it be justified in a time of ecological urgency?

As a college president, I would often address prospective students and families. Why should they consider the environmental field? And in other circumstances (with colleagues, friends, or in public settings) I find myself explaining the virtues of an environmental career and life, or how to incorporate a sustainability ethos into one's life practice. The essence of my appeal is twofold. I explain that environmental sustainability is the ultimate service profession. Wherever you are, however you work, you are engaged in activity that serves your neighborhood, community, and planet. Service is rewarding, engaging, and meaningful. Second, by studying sustainability and the natural world, you are gaining a deeper understanding of life processes. In so doing, you are constantly reminded of the mystery and wonder of the biosphere. As you do so, you gain an appreciation for the sanctity of life.

I can think of no better way to integrate personal growth and the pursuit of a career. The justification is embedded in this appeal. Thoreau's daily practice of observing nature was far more than a testimony to direct experience. It was a way to build appreciation for the very circumstances of his life. Rather than take the natural world for granted, he chose to probe its intricacies. In deepening appreciation, he summoned gratitude. The good life beckons gratitude. For Thoreau, this is the very essence of human flourishing.

How can this sensibility be relevant to the 24/7 world of contemporary higher education? It isn't easy. Expressions of gratitude can be washed away in cynicism, sarcasm, anxiety, and stress. Or they may be perceived as sanctimonious. How can I express gratitude when you have just slashed my budget? The budget-cutting mentality, the trappings of accountability and assessment, the constant need to justify higher learning beyond sheer productivity and career building—these pressures can shatter gratitude into the scattered fragments of spare change. Where does Thoreauvian simplicity belong here?

Perhaps the most vivid reminder of gratitude is to call attention to the great privilege of education itself. Just as we often feel entitled to the earth's bounty, so do we expect the provision of a good education. Yet the great majority of the world's population has no access to either. These two fundamental expectations—the fruits of the earth and the gifts of higher learning—are indeed the culmination of the good life, and taking them for granted leads to their squander. Budget cutting is so threatening because it ultimately implies less access to both prospects. Let us be thankful for what we have and conserve its best use.

The idea of gratitude is at the heart of Thoreauvian simplicity. It is also the very essence of the sustainability ethos because it teaches that the culmination of gratitude is reciprocation. Reciprocation implies giving back what you have received. It involves exchange, transformation, and acknowledgment. Reciprocation is a circulation from the biosphere through human awareness and back again, passing through social networks, educational venues, creative expression, and community service. It is the very foundation of human flourishing. If reciprocation and gratitude are essential to the good life, how can such qualities become intrinsic to the curriculum of higher education?

7

Curriculum

Too Many Majors

The first time I ever looked at a college's course catalog was in the mid 1960s. I was thoroughly bored with high school and I couldn't wait to explore what colleges had to offer. I received many catalogs in the mail. I loved flipping through the pages, contemplating the courses I might take, the majors I could pursue, and the books I would read. Those catalogs seemed to embody all of the world's knowledge, and I was thrilled at the prospect of having access to it.

Many years later, when I was teaching environmental studies programs and courses, I wrote some copy for a college's catalog, assuming that it was a great way to make a compelling case for the importance and excitement of the environmental field. I took great pleasure in curricular design, using the catalog copy as an evocative means to explore engaging ideas about teaching and learning. When I worked at Antioch New England Graduate School, we were designing and building new programs. The catalog had to distinguish and justify our course offerings as well as our philosophy of learning. It had to provide our university with a recruitment edge.

When I arrived at Unity College in 2006, times had changed. Catalogs were replaced by view books. Facebook, websites, chat rooms, and videos were the preferred recruitment tools. Catalogs served a very different purpose. Nevertheless, students were interested to know what they would be studying, what programs our college had to offer, and what kinds of courses were available. And when I was a candidate for the Unity College presidency, I was a bit like a prospective student, or at least I tried to put myself in those shoes. So I thoroughly reviewed the course offerings, majors, and programs. Surely the search committee would ask me what I thought about the curriculum and assess my views on how the curriculum might change.

What I found was surprising. The college had almost as many majors as it had faculty members. Some of the majors sounded so similar that even as a professor of environmental studies I had a hard time distinguishing them. They didn't seem to present a unifying vision, a clear statement about the future of the field, and I couldn't imagine how they could help prospective students understand the strengths of the college. I understood how the majors evolved over time, reflecting the specific interests of particular faculty members, but it was clear that change was needed. I suggested to the search committee that the curriculum be overhauled, emphasizing that it should anticipate the future job markets for environmental studies and sustainability, should be reorganized around no more than a half-dozen academic strengths of the college, and should dramatically reduce the number of majors so as to conform to those strengths while assessing the future of the field. I also wondered about the lack of a major or a program in sustainability. Finally, I explained that the curriculum should be evocative and visionary, as it was one of the college's most essential recruiting tools. Once the programs were redesigned, they should be explained attractively and marketed and "branded" well, and the admissions people should work closely with faculty members to ensure clarity and consistency. At that point I was merely a candidate for the presidency, and I could be free with my consulting advice.

And then I got the job. It is one thing to advise a search committee on what a college should do; it is another thing altogether to be responsible for implementing that advice. In my first year I mainly gathered information, learned about the college's academic strengths and weaknesses, and initiated a search for a new vice president for academic affairs (VPAA). I worked quickly with the very helpful interim provost to install two new sustainability majors. We felt that those were necessary for the college, and that any delays would set us even further behind in recruiting interested students.

At the beginning of my second year as president, I asked the new VPAA to organize a two-year academic planning process. It would include gathering data from nationally renowned experts in environmental studies and sustainability, a thorough assessment of Unity College's interests and strengths, and an investigation of the programs that were most attractive to prospective students, and would culminate in five "centers" and a dozen majors. I told the VPAA and the faculty that, insofar as they were stewards of the curriculum, the academic plan was their project and responsibility, and that they would have the authority and autonomy to implement it. I would be available for consultation, but I would participate in the planning meetings only on their invitation. I urged them to emphasize innovation, creativity, and change. Once the process was complete, I would try to find funding to implement the changes. As someone who specialized in designing environmental studies programs, it was very difficult for me to stay on the sidelines. But I knew that the best I could do was offer timelines and guidelines, because ultimately the faculty would be happy only if they managed the process.

The planning process proceeded in fits and starts. There were plenty of arguments and difficulties. It took slightly longer than two years to get it done. Coordinating the new "centers" and "majors" with enrollment management, the registrar, and all corners of the college was time consuming and challenging. Yet when the process was finally accomplished, the faculty and the staff were pleased with the outcome. Five centers (Biodiversity, Environmental Arts and Humanities, Experiential and Environmental Education, Natural Resource Management and Protection, Sustainability and Global Change) were established to house eighteen well-defined majors.[1] The focus was on the three academic strengths of the college: science, service, and sustainability. I, too, was very pleased with the outcome. We had successfully revised the curriculum, and sustainability became an academic foundation for the college.

The academic master plan was just half of the picture. For sustainability to have a major conceptual effect, it had to have co-curricular implications. Could every campus sustainability initiative have an educational component? I wanted all of our students to "learn by doing." I considered all visitors to campus as potential learners. (See chapter 8.) My aspiration was that the college's curricula reflect a transparent relationship between classroom study and the routines and behaviors of everyday life. I thought using "science, service, and sustainability" as a slogan was a fine way to weave those routines through campus life and study. These changes, too, would require the systematic alignment of all campus constituencies. Students, staff members, faculty members, members of the senior leadership, alumni, and board members would have to work together.

I recognize that Unity College's challenge may seem to have been an easy one, and that it may not be relevant to other campuses. After all, Unity had already described itself as "America's Environmental College." Surely it couldn't have been terribly difficult to revitalize the curriculum so that sustainability become primary. Still, there were doubters, blockers, and curmudgeons. Curricular change was a source of constant tension and anxiety. It was never easy, and at times I wanted to throw my hands up in frustration. The faculty members were already overworked, teaching seven courses in two semesters. They had every reason to balk at this new work assignment. But they didn't. And we accomplished our goals. Why? Ultimately the faculty understood that the curriculum revitalization, linked to sustainability, was in the best interests of the college, and would enable them to construct their own academic future.

The first sections of this chapter cover some of the political dynamics that affect curricular change, especially as related to sustainability. Who are the agents of curricular change, and how can they maximize their effectiveness? Then I'll discuss the necessity of a co-curricular orientation. Why is "learning by doing" a necessary framework for sustainability? How might we envision the campus as a sustainability design studio? This leads to some of my recommendations for an adaptive sustainability curriculum—techniques that may work in a variety of institutional settings. Readers should remember that a sustainability curriculum is a response to a planetary challenge, a means to mobilize solutions and constructive change, and an opportunity to use the campus as an educational venue for doing so.

Curricular Transformation

A college's curriculum reveals its most deeply embedded values. Accordingly, it represents the hopes, dreams, expectations, and aspirations of all campus constituencies. The best way to understand a college's aspirations for its learners is to survey its courses, programs, and syllabi. The curriculum, although typically vested as the responsibility of the faculty, is also influenced by all interested parties—donors, governments, alumni, and accrediting bodies, as well as students and their families.

A curriculum is rigorously scrutinized as a repository of values. Consequently, it is perceived as the most likely platform for educational reform, but also (because of all the parties involved) where such reform is least likely to occur. Its potential as an agent of educational change agent is obvious. Where else can you so thoroughly engage the hearts and minds of the students and the faculty? However, the curriculum also "reproduces" (to use C. A. Bowers' term) the historical and cultural assumptions that are built into the very structure of university life.[2] With so much at stake, and so many interested parties, inevitably there are layers of approval processes and lengthy discussions. These often lessen the prospects for change.

The sustainability ethos calls attention to what we learn, how we live, and how we respond to a planetary environmental challenge. Its practitioners seek to influence and ultimately transform the curriculum. They correctly perceive the curriculum as crucial to educational change. At any workshop, conference, or gathering of sustainability professionals, they probably will be deliberating on their strategic options for moving the sustainability curricular agenda forward on their campus. As a participant in countless sustainability-related meetings, conferences, retreats, and strategy sessions, and having served on the steering committees of many organizations, I find that a few prevailing themes repeat themselves.

In any discussion of sustainability in relation to a college's curriculum, someone will make a compelling case for using sustainability as a means to transform the curriculum. The argument goes something like this: Climate change, environmental pollution, and loss of biodiversity are happening much more rapidly than anyone anticipated. These changes will dramatically and drastically affect the quality of human life. Only a profound transformation of values will enable us to address this challenge. Hence, we have to go right to work on the curriculum so that every college graduate understands, internalizes, and cares about these issues. This curricular transformation must be comprehensive and swift, encompassing all aspects of teaching and learning, from foundation courses for freshmen to the professional programs. It must emphasize interdisciplinary, systems-oriented thinking. The campus's infrastructure should set an example for the ideas we are teaching.[3] I have heard this argument in one form or another, as related to environmental studies and/or sustainability, since the 1970s.

A sympathetic critic, often someone in senior leadership, may respond as follows: "There are many constituencies on our campus. There are people who emphasize the importance of international learning and diversity, others who devote their work to gender and sexual preference studies, service learning or civic engagement advocates, and dozens of others who demand attention, preference, and urgency for their agendas, too. Then we have to deal with other curricular advocates, ranging from those who bemoan the decline of basic study skills to those who wish to resurrect the humanities. By virtue of their training and scholarship, all of these constituencies have a stake in the curriculum. On the top of that, our campus is suffering from unprecedented budget cuts. Our students are less interested in values and more concerned about careers and jobs. We care deeply about sustainability, but it is hard to see how it can have precedence in these times."

The advocates of transformation will then suggest that the role of senior leadership, especially the president, is to boldly articulate how and why sustainability is necessarily the foundation for all curricular issues, as it broadens the most fundamental awareness—the inextricable link between the quality of human life and the health of the biosphere. Without such an emphasis, colleges and universities are abnegating their leadership responsibility. Further, we can use the institution's financial challenges as a means to ask whether the curriculum is sufficiently pertinent, vital, and responsive. Which of our programs are necessary? It is understandable that the economy is affecting the priorities of the campus. Let's incorporate the concepts of opportunity, preparation, and innovation into the language of sustainability so as to demonstrate its utility as a unifying mission.[4]

This conversation reiterates the political necessity of curricular debate, the role of leadership in moving the conversation forward, and the heightened stakes when words such as "transformation" are thrown around. It explains why advocates of sustainability perceive themselves as agents of curricular change and why their work is often controversial. It is a daunting task to transform a curriculum. The process of curricular transformation is implicit (and often explicit) in the mission and operating agenda of the campus sustainability movement.

Benjamin Barber, in his classic work *Strong Democracy*, describes agenda setting as one of the nine functions of strong democratic talk. Two observations stand out. "He who controls the agenda—if only its wording—controls the outcome." Who gets to speak and when do they get to do so? "The ordering of alternatives can affect the patterns of choice as decisively as their formulation."[5] For a sustainability curriculum to be fundamental to the academic mission of a campus, it must be prominently discussed in multiple departmental settings.[6]

Strategic Curricular Innovation

As I noted earlier, I have spent my entire career in settings where curricular innovation was crucial to the success of the college. In order to attract students, generate the necessary revenue, and compete in the higher-education marketplace, we had to design and develop interesting, innovative, and career-oriented programs that could successfully recruit, retain, graduate, and place our students. This is especially true of, although not exclusive to, the environmental and sustainability fields, as innovative and timely programs are the first to capture the attention of prospective students and funders. Additionally, the field (since the late 1960s) continues to emerge and evolve, so the curriculum has to adapt to changing circumstances and new knowledge. Program design has necessarily entailed a curricular process of adaptive change. In these settings, the faculty members not only have the responsibility of designing the curriculum, but they must also be aware of market trends, emerging professional needs, peer program developments, and enrollment management scenarios. The faculty is necessarily engaged in all aspects of program design, planning, and development. The great advantage of this is that they hold the keys to their own future and have great autonomy and support for curricular design. The disadvantage is that they have less time for research, writing, and scholarship. This requires a delicate balancing act, especially regarding relationships between the faculty and the administration.

In higher education's new era of austerity, colleges and universities that engage their faculty members in all aspects of program planning, development, and recruitment are more likely to cultivate the market awareness that will ensure their survival. This is a controversial position. Although traditionally faculty members have always been curricular stewards, they haven't necessarily been as engaged in matters of enrollment, marketing, strategic planning, and cost accounting. This is now changing in a variety of institutions, not just at smaller, independent institutions whose revenues come largely from tuition. Schools with innovative faculty members who are adaptive, flexible, and anticipatory, and who have the capacity and willingness to implement creative new programs, will gain a strategic advantage in the higher-education marketplace. This is true for most fields and disciplines, but especially crucial for the sustainability field.

The sustainability curricular agenda has multiple strategic considerations. To be successful, it must address the specific strengths, mission, and circumstances of the institution. What are the inherent academic strengths of a campus? What are the academic priorities of the campus? What are the core values of the institution as defined in the mission statement? What programs will attract and retain students? How does sustainability enhance, refine, challenge, and amplify those strengths? For example, in difficult economic times, campuses will emphasize work preparation, career development, and economical renewal. How shall these priorities integrate a sustainability curriculum?

The singular contribution of the North American higher-education system is the extraordinary diversity of types of campuses and the amazing variety of curricular approaches. This necessarily informs the sustainability curricular agenda. Every campus will choose a uniquely interesting path, informed by its educational philosophy and setting. The sustainability ethos will necessarily challenge that philosophy, but it need not resort to political correctness, universal mandates, or judgmental pronouncements in doing so. Rather, the faculty (with encouragement from the senior leadership) can reinterpret sustainability as it best enhances the institution's academic strengths and weaknesses.

Senior leaders should tread carefully in curricular matters. They can provide a vision, a concept, a challenge, and a strategic timetable, but ultimately the faculty and the students will have to "own" the curriculum if it is to be successful. The leaders should encourage and reward innovation, challenge the status quo, serve as institutional change agents, position the campus as a curricular exemplar, provide support, connect faculty members with prominent external constituents, and assert the necessity of promoting curricular change as a process of adaptive management.

Sustainability and Curricular Politics

It takes a long time to transform a curriculum. If you view the "transformation process" as a race against time, you may not win. Depending on the institution's size, its culture, and its readiness to change, you will encounter dozens of approval processes (e.g., for requirements, majors, programs, or courses). Some of these processes are inclusive and some are unilateral. Often a recalcitrant administrator or a reluctant group of faculty members can impede progress. Then there are the accrediting bodies. It seems that every curricular approval decision eventually requires deliberation, consultation, and justification. These processes take time. And I'm not sure it can happen in any other way in settings that ostensibly value, inclusion, conversation, and transparency, or when bureaucracies are involved.

Of course, a president or a provost can always try to mandate change by offering directives, rewards, or other formalized incentives. These can backfire too. Many college presidents and provosts have learned this the hard way. Despite their best intentions, their curricular mandates may not be as widely shared as they think. Even if they are, if they proceed without adequate participation, their initiatives may fail. College campuses are not the best places for unilateral proclamations.

Such are the realities of campus curricular politics. They are sources of numerous frustrations. Discussions about the slow pace of change also abound at various sustainability gatherings. This frustrates senior leadership as well. I have been in many discussions among college presidents when they bemoan the slow pace of the faculty's readiness to change, or the challenges they face with their governing boards. Faculty members often complain about senior administrators who don't sufficiently support their work, or other faculty members who are unwilling to change.

I cite these frustrations simply to reiterate the numerous constraints on curricular transformation—constraints that are almost always linked to campus politics. Yet sustainability curricula are proliferating. How is this happening? First, presidential leadership initiatives (despite the caveats described above) can have a huge effect on a curriculum. Second, there are emergent, grass-roots processes that develop in the interstices of campus politics, influencing a curriculum in both subtle and direct ways.

Some university presidents are taking bold leadership steps to promote comprehensive sustainability initiatives. This raises important governance challenges (see chapter 4), requiring finesse and flexibility. A primary role of a president is to articulate a shared vision, adapt it to the cultural context of the campus, delegate effective change agents, and provide rewards and incentives accordingly. It is much more effective when the entire campus is engaged in sustainability initiatives, when all involved see their efforts as intrinsic to the campus's identity, when the school incorporates them in its literature and its promotional materials, and when these efforts are embedded in the curricular substrate of everyday learning. If the president (as the "educator in chief") supports these efforts, they are more likely to penetrate the curriculum. The colleges and universities whose presidents are engaged with on-campus sustainability efforts are the ones at which robust curricular changes are more likely.[7] Another influential approach involves using networks of chief academic officers to discuss the implementation of curricular changes.

There are also processes that seem to transcend campus politics, at least initially. Students, faculty members, and staff members often promote grass-roots sustainability initiatives that have curricular implications. Many campuses have sustainability learning communities, freshman seminars, more advanced courses, and/or speaker series that eventually develop into new programs and majors. Sometimes the popularity of these efforts convinces administrators of their value by generating revenue, attention, and interest. Student Services departments have an increasingly important teaching role on campus. They design orientations, campus events, residence-hall gatherings, and other programs that directly affect everyone on campus. In some settings, the Student Services department works with the faculty to plan seminars, outings, and other educational experiences. There are nationwide networks that organize campus sustainability activism, and many of them include large numbers of students.[8] Many campuses link these conferences to courses and the curriculum. Students who attend these conferences often present papers or posters about their activities. In some cases, students and faculty members share an avocational interest in sustainability that they apply to campus projects. All these efforts are enhanced and amplified through social media. These sustainability projects have curricular implications, yet they sidestep the approval processes that often inhibit curricular change.

Co-Curricular Sustainability

When (as a consultant) I meet with a campus sustainability team, the members typically describe the "co-curricular activities" they have launched, emphasizing learning opportunities that supplement or support the traditional classroom offerings. Among the co-curricular activities I have had described to me are a community-based electronic waste recycling center, the use of native plants for campus landscaping, a cafeteria's composting center, a community wind assessment service, an organic gardening collaborative, and a sustainable living residence hall.[9] A common feature of such activities is that they bring together students, staffers, faculty members, and members of the outside community. Such projects have enormous educational value because they are closely linked to the everyday life experiences of their participants. They also generate collaborative learning. The facilities staff may teach landscaping techniques to students. A student may explain to a faculty member how her dormitory's solar panels work. An expert gardener may explain to the facilities staff what native plants are best suited for the campus.

It is instructive to consider how sustainability initiatives have influenced American college and university landscapes since the year 2000. The great majority of the projects cited above have emerged since then. Hundreds of thousands of students, staffers, and faculty members have been exposed to, learned about, or participated in these projects. There is much greater awareness of energy conservation, organic food and nutrition, recycling and materials use, housing, bicycling, and other aspects of sustainable campus life. Many of these initiatives were developed, without any formal curricular approval process, in response to student interest, administrative planning, community involvement, and the impact of the campus sustainability movement.

On some campuses, co-curricular projects are neatly coordinated with faculty research, classroom assignments, internships, work-study jobs, or the routines and habits of everyday life. For example, architecture colleges design LEED buildings, business schools organize "green profit-making centers," engineering colleges coordinate research on solar energy designs, and agricultural programs study experimental organic food growing. From environmental art to climate adaptation, from social behavior research to aviation engineering, the reach and breadth of sustainability-related programs is extraordinary. I have attended many faculty sustainability workshops. I am always surprised and delighted at the variety of subject matter backgrounds. Although historically much sustainability curriculum has grown out of environmental studies, that is no longer the case. An environmental sensibility still informs how we think about sustainability, but the academic perspectives now range from business to studio art to engineering, and to every conceivable place beyond and in between.

Two encouraging features of co-curricular projects are the extent to which these programmatic initiatives inspire interesting new courses and majors. There is a new generation of sustainability academic programs that are experientially oriented, multi-disciplinary, and practically engaged, and that utilize the campus and nearby community for much of their research.[10] Many of these programs are just getting started. In the years to come, we will have a much better evaluative sense of their effectiveness.

On many campuses, even those with these great initiatives, advocates of sustainability often comment that sustainability efforts are disparate and far-flung, lacking coordination, and systematic integration and express concern that the new programs will only serve a small percentage of the student body. Transformational change requires that every student be thoroughly exposed to the principles of sustainability science and that a sustainability ethos inform all campus activities. I have visited many campuses, from small liberal arts colleges to large urban universities, where members of the sustainability team may be proud of their accomplishments but skeptical that their efforts are sufficiently comprehensive, or that they have moved beyond "preaching to the choir."

For all of the reasons mentioned earlier, campus curricular politics can inhibit such transformational change. Such frustration is understandable. Yet by any objective measure the co-curricular initiatives are moving quickly, and their impact is undeniable. The research regarding sustainable behavior is still in its early days (it will take several decades of longitudinal study), and much of the current discussion involves how to foster sustainable behavior change.[11] Until more research has been conducted, we can't say definitively that a college student who spends four years living in a "green" residence hall will forever live her life according to the principles of sustainability. However, common sense suggests that the routines and daily practices of sustainable living may raise awareness about basic quality of life issues. That's the motivating rationale behind co-curricular initiatives of all types and certainly for sustainability practices. Academic and co-curricular programs have a reciprocal and synergistic relationship. That's why campus sustainability efforts are most effective when they work in both domains. It may take too much time to transform a campus curriculum, but in the meantime there is much to be accomplished through "learning by doing."

Learning by Doing

Throughout the twentieth century, theorists (most notably John Dewey) wrote elaborate texts espousing the importance of "learning by doing." More recently, constructivist learning theory has emphasized the cognitive necessity of "providing learners with the opportunity to interact with sensory data and construct their own world."

In a superb article on constructivist learning theory, the museum educator George Hein outlines the nine basic premises of a constructivist approach: (1) Learning is an active process in which the learner uses sensory input and constructs meaning out of it. (2) People learn to learn as they learn. (3) The crucial action of constructing meaning is mental. (4) Learning involves language: the language we use influences learning. (5) Learning is a social activity. (6) Learning is contextual. (7) One needs knowledge to learn. (8) It takes time to learn. (9) Motivation is a key component to learning.[12] This is a common-sense formula for just about any subject, generation, or culture. It is a deeper, research-based elaboration of the "learning by doing" concept. And yet much of formal education fails to implement this common-sense approach. Much ado is made about how computer screens increasingly dominate the learning process, and depending on your values this is a cause for opportunity or concern. But what makes computer learning attractive? It isn't the screens themselves. Rather, it is the fact that computer and video games, social media, and other uses of information technology provide an instantaneous manipulation of variables with rapid response. Computer games and social networks, especially linked together, embody the nine principles listed above. One reason they are compelling is that formal education doesn't provide sufficient "learning by doing."

How is all this relevant to a sustainability curriculum? The sustainability ethos embodies life skills as well as theoretical concepts. Much of the sustainability literature preaches the integration of the two. There is no point in sitting through lectures and labs about energy efficiency if the campus itself isn't energy efficient, or if its residents don't live and work in buildings that exemplify those lessons. What purpose does it serve to discuss nutrition or alternative food-production systems if the campus doesn't serve and/or grow local and organic food? How can you teach ecology if your students never get engaged in field-based natural history?

What you know and how you think always reveal how you live. Students are quick to point out the inherent disconnection that results when this principle is neglected. When students (especially those between the ages of 18 and 22 years) come to a college campus, many of them have just left home for the first time. They are experimenting and exploring, wondering about their values, who their friends will be, and how they should live their lives. They are constructing their social identities. They are establishing behaviors and practices (beyond studying) that will become the basis of lifelong living and learning. This is also true (in different ways) for adult learners. For many of them, the campus experience will be a watershed life experience. That's why it is so important that campuses provide tangible, visceral sustainability experiences—from recycling to energy efficiency—that are integrated with the daily lives of the students, whether they are residents or commuters. The best sustainability curriculum is one that provides the hands-on experience of living, implementing, and designing a sustainable campus, tangibly linked to the more formal curricular expectations of programs and majors. The experience of sustainability practice is the very best teacher. Many advocates of sustainability urge campuses to become "living laboratories" of sustainability practice. In the next section, I'll address the curricular implications of this idea.

The Campus as a Sustainability Design Studio

What does it mean to conceive of the campus as a sustainability design studio? The idea of design is derived from art, architecture, planning, engineering, fashion and gaming. It suggests a creative and detailed planning process, organized to achieve a set of specified outcomes. "Sustainability" implies a suite of values—ecological criteria as manifested in interactions between humans and nature are embedded in the design criteria and process. "Studio" connotes a workspace, a place where individuals and groups can mess with materials, use their imagination, invent new schemes, and try things out. Consider the campus as an improvisational sustainability playground. More specifically, I am advocating that everyday life practices, work and service projects, research initiatives and innovations, campus planning and community partnerships contribute to sustainable design and curricular initiatives. There are four components to this "studio" process: landscape, planning, research, and behavior.

Every college or university is a physical landscape with an ecological setting. Its history and culture are reflected in its buildings, its grounds, and its curriculum. A campus's landscape makes a profound impression on everyone who uses it. We admire beautiful campuses because they inspire us. Imagine the educational potential of a campus whose landscape is totally geared toward exploring sustainability—transparent and innovative use of local or recycled materials, building designs and retrofits that reflect a visible commitment to conservation, edible landscaping and gardens interspersed on grassy lawns or urban street corners, or dormitories with rooftop gardens. What a great way to involve students (and their families), staff, and faculty members in learning about sustainability through their daily life routines and habits!

All of a campus's master planning processes should emphasize sustainable design and involve as many participants as is feasible. There is no better way for a community to learn about sustainability than to design the buildings, the layout, and the landscapes of their living and working spaces. A good master planning process incorporates the values, finances, and mission of the institution while leaving sufficient room for unanticipated circumstances. Flexibility and improvisation are crucial. If the planning process allows for the campus to become a sustainability design studio, it builds sustainability criteria into all aspects of the evolving and emerging campus landscape.

Planning a campus landscape entails data gathering and research. Every campus sustainability initiative can be seen as an educational research experiment. We are fortunate to be living in a time when dozens of imaginative, technical, and interesting sustainability solutions are being proposed. Which of them are most appropriate for a campus at any given time? How do we determine this? Who is taking notes and gathering data? How might a campus experiment with these ideas? There are many ways that colleges and universities are involved in sustainability research. Programs in materials science, renewable energy, ecological architecture, organic agriculture, urban policy, ecological economics, and environmental perception, as well as other subjects can contribute to campus sustainability initiatives.

The most compelling lessons in life come from ubiquitous daily routines and behaviors. Our common habits (for better or worse) are the building blocks of a comprehensive worldview. When you disrupt those routines with different expectations and structures, you are likely to engender deep learning, and then, in turn, build new routines and behaviors. A sustainability curriculum is empty if it is primarily theoretical. However, it is shallow if it lacks study, reflection, and substance. Students watch what we do and how we think. That's why work colleges[13] have such an extraordinary effect on their students, why many schools are now emphasizing experiential learning, and why service learning has become an important aspect of higher education.

Social Capital and Ecological Intelligence

Another reason for turning the campus into a sustainability design studio is to cultivate social capital and ecological intelligence. Conceive of the studio as the canvas, social capital as the process, and ecological intelligence as the outcome. Integrating these concepts builds a resilient cultural commons. Let's explore these relationships by looking more closely at these terms, why they are important concepts for a sustainability curriculum, and why their synergy is compelling.

In chapter 5, I referred to Feldstein and Putnam's definition of social capital "as the collective value of all 'social networks' (who people know) and the inclinations that arise from these networks to do things for each other ('norms of reciprocity')." By turning the campus into a design studio, we construct a campus-wide dialogue. The studio creates a public experience. It encompasses networks of expertise and deliberation. A successful sustainability initiative should strive to build social capital. Learning how to use social capital should be a foundation for any sustainability curriculum.

Building social capital is a prerequisite for creating an enduring respect for the cultural commons. The educational theorist C. A. Bowers defines the cultural commons as "the intergenerational knowledge, skills, and mentoring relationships that enable members to be more self-reliant in the areas of food, healing, creative arts, craft skills, narratives, ceremonies, civil liberties, and other aspects of daily life that are less dependent upon consumerism and participation in a money economy." Bowers insightfully suggests that the commons concept broadens our traditional notions of community to include ecological and cultural criteria, thus expanding the spatial and temporal boundaries of our considerations. In this regard, the sustainability design studio may revolve around a planning scenario spanning 25–50 years but should encompass a more penetrating view of the campus in ecological and cultural space and time. Bowers uses the term "ecological intelligence" to convey the "importance of learning the cultural patterns of moral reciprocity essential to community—while also retaining the more contemporary understanding of the behavior of natural systems as ecologies."[14]

"Ecological intelligence" is Daniel Goleman's conception of how to build learning capacity to assess the relationship between everyday life decisions and the biosphere. "*Ecological*," Goleman writes, "refers to an understanding of organisms and their ecosystems, and *intelligence* lends the capacity to learn from experience and deal effectively with our environment. Ecological intelligence lets us apply what we learn about how human activity impinges on ecosystems so as to do less harm and once again to live sustainably in our niche—these days the entire planet."[15]

Let me reiterate this curricular sequence. The sustainability design studio is a public venue for integrating landscape, planning, research, and behavior. The design process coordinates individual effort and collective aspiration, utilizing a "learning by doing" approach. Learning how to maximize social capital takes full advantage of shared expertise and prepares participants to fully engage in collaborative working and living environments. Understanding the cultural commons illuminates the intergenerational setting for sustainability initiatives. The ultimate outcome of this process is enhanced ecological intelligence.

An Adaptive Sustainability Curriculum

There are countless approaches to constructing sustainability curricula. There are many outstanding guides, texts, syllabi, and programs for doing so. Whatever your interest might be—perhaps a freshman course introducing sustainability concepts, different iterations of sustainability majors, a comprehensive reworking of an entire liberal arts program, or sustainability courses for professional schools—you can be sure that someone has probably launched a good effort.[16] Educators have been proposing such curriculum for more than forty years.[17]

Curriculum development is a dynamic, situational, and participatory process. That's why so many of these good efforts repeat themselves. We often hear about the inefficiency of "reinventing the wheel" when it comes to curriculum. It is obviously useful to be aware of what others are doing, and to be inspired by great ideas. As stewards of the curriculum, most faculties aspire to incorporate their expertise and values as essential to the teaching and learning process. Hence there are many "home-grown" curricular efforts. In view of these contingencies, I am proposing adaptive curricular guidelines that may be relevant in diverse institutional settings, acknowledging, too, the rapidly changing external environment that brings new knowledge and situations to bear on academic preparation.

The ultimate outcome of a sustainability curriculum is to enhance individual and collective capacity to implement the sustainability ethos in diverse personal and professional settings. An understanding of sustainability principles requires deepening ecological intelligence. Enhancing social capital strengthens the implementation capacity. Both processes are best achieved with a "learning by doing" emphasis. The formal curriculum is supported with co-curricular initiatives. The campus is a sustainability design studio serving as a laboratory and practice field.

Informed by this configuration, my impressions of the sustainability field and its future, and my educational philosophy, I propose four broad categories as a curricular foundation: biosphere studies, social networking and organizational change, the creative imagination, and sustainability life skills. These are mutually reinforcing and reciprocal categories. They correspond with the classic formulation of natural and physical science, social science, the humanities, and professional practice. However, they are reconstructed to emphasize sustainability for the twenty-first-century learner. For each foundation, I'll briefly present a core learning process, distinguish a personal and public dimension, and then suggest some areas for substantive inquiry and experimentation. These are intended as ways of stimulating discussion, controversy, and inquiry for constructing a sustainability curricular agenda. Consider them as a catalog of emerging curricular design potentials. In an undergraduate curriculum, these categories may be conceived as a way to revitalize the traditional liberal arts. Or any one category can serve as the foundation for majors. The specific examples for study can be specific courses, research projects, or graduate programs.

Biosphere studies emphasize an understanding of earth-system processes and global environmental change. The challenge is how to develop a conceptual sequence that helps students perceive, recognize, classify, detect, and interpret biospheric patterns, and to do so through local ecological observations. Elsewhere I describe this as "pattern-based environmental learning."[18] The purpose is to better understand and internalize global environmental change. The personal dimension involves the development of natural history observation skills so as to enhance appreciation of home and habitat. The public dimension involves how to contribute those observations and assessments to global networks of biospheric data collection. Examples for study may include biogeochemical cycles, evolutionary ecology, restoration ecology, watersheds and fluvial geomorphology, biogeographical change (species migrations, radiations, and convergences), plate tectonics, and climate change.

Social networking and organizational change describes how to enhance, cultivate, and utilize social capital. This includes a personal dimension—providing students with the ability to better understand how they learn and think, how they respond to stress, and how to maximize psychological and physical wellness. The public dimension promotes the ability to interpret collective behavior in organizational settings. The learning process involves how to integrate the personal and social dimension so as to maximize human flourishing in diverse institutional settings. Examples for study include cognitive theory, neuropsychology, organizational process, change management, behavioral economics, ecological economics, social entrepreneurship, decision-making science, adaptive management, and social networking theory.

The *creative imagination* entails the cultivation of an aesthetic voice, personal expression, and improvisational excellence to enhance the arts, music, dance, play, literary narrative, and philosophical inquiry. The personal dimension emphasizes how to use the creative process as a means to explore questions of ethics, meaning, and purpose, how to maximize aesthetic joy, and how to express emotional responses to challenging sustainability issues. The public dimension develops the capacity for collective expression in social milieus—how to use public spaces (buildings, parks, campuses, and so on) to promote learning about sustainability, or to effectively construct public design studios. Examples for study include environmental art and music, acoustic ecology and sound design, biophilic design and architecture, environmental interpretation, environmental perception, environmental ethics, ecological identity, the aesthetics and epistemology of patterns, game design, information design, and biomimcry.

Sustainability life skills is the application of sustainability principles to the routines, behaviors, and practices of everyday life. The personal dimension involves the individual behaviors of sustenance, shelter, transportation, health and domestic life. Further, it emphasizes how to incorporate sustainability principles into one's career and professional choices. The public dimension involves how to support organizational or regional sustainability efforts, including procurement, ecological cost accounting, recycling, health services, and/or other forms of community capacity building for sustainability. Examples for study include organic agriculture, nutrition, home building and engineering, construction, alternative energy, energy and water conservation, waste management, alternative transportation, sustainability metrics, habitat restoration, gardening, urban and regional planning, career development, reflective practice, and service learning.

One of the challenging and exciting aspects of curriculum development is how to keep up with rapidly changing new knowledge. In view of the proliferation of ideas and research, the emergence of rapid information dissemination systems, and the extraordinary levels of specialization and/or hybridization of knowledge, it may be impossible for curriculum to keep pace with new developments. A good way to finesse this challenge is for curriculum designers to emphasize the importance of critical, creative, and cognitive thinking skills and approaches. And yet the fields of cognitive psychology and neuroscience (which ultimately inform how we think about thinking) are also in rapid transformation. Since much of education concerns learning how to think, there should be an active conduit between the latest research in these realms and the hands-on process of curriculum development. By way of example, and just to highlight the pertinence and value of this research, consider the ramifications of the following projects.

In *Thinking, Fast and Slow*, Daniel Kahneman describes his lifelong research in the cognitive psychology of human decision making. Kahneman is mainly interested in "the biases of intuition" and how their understanding may "improve the ability to identify and understand errors of judgment and choice, in others and eventually in ourselves, by providing a richer and more precise language to discuss them." He is concerned that "many people are overconfident, prone to place too much faith in their intuitions."[19] His book is a detailed, accessible, and practical assessment of how to overcome such biases so as to improve decision making and ultimately promote a sense of well-being.

In *The Emotional Life of the Brain*, Richard Davidson recounts a lifetime of research in "affective neuroscience, the study of the brain mechanisms that underlie our emotions and the search for ways to enhance people's sense of well-being and promote positive qualities of mind." Davidson elaborates on six emotional styles (resilience, outlook, social intuition, self awareness, sensitivity to context, and emotion) and demonstrates how they are linked to a neural signature—how they correspond to neural pathways in the brain. He is particularly interested in neuroplasticity, the idea that with practice you can strengthen and alter corresponding neural pathways. His research on meditators, for example, "has shown that mental training can alter patterns of activity in the brain to strengthen empathy, compassion, optimism and a sense of well-being."[20]

While reading the above-mentioned works (and others),[21] I was struck by the possible curricular applications. All these works are replete with tangible exercises, experiments, and experiences that help the reader understand the basis of a range of psychological challenges, from decision making to perception to stress reduction. Wonderful courses and programs can flow from this material.

An improved understanding of decision making behavior is of great interest for the sustainability curricular agenda because it illuminates why we make choices. Why do we make choices that don't necessarily serve our self-interest or the collective interest, or that damage the ecosystem on which we rely? How do we enhance the clarity of decision making and thus promote human flourishing in the biosphere?

I'll conclude this section with a comment about requirements, majors, and programs and their relationship to curriculum development. Even in the most progressive institutions, debates about these issues can be vigorous and stultifying. Requirements can sometimes be overrated as if they are rules of law. Curricular change is most often hung up here. From a substantive perspective, faculty members have vested interests in what they perceive as the essential preparatory sequences. Administrators tend to favor cost-effective efficiencies. Some of the most intractable campus controversies occur at the intersection of these approaches. Here is where budgetary issues most often dictate academic compromise. Knowing full well the complexities of this type of challenge, I advocate curricular flexibility when it comes to requirements. Preparatory sequences are not necessarily as logical as the faculty perceives them to be. And standardized approaches rarely meet the long-term needs of either students or faculty members. In my view, requirements and programs should serve as recommendations—strategic travel guides—for how to proceed through a terrain of courses and opportunities. A student wants guidance in navigating a seemingly intricate variety of options. The faculty wants to provide the most consistent framework. A sustainability curriculum will best proceed as a process of flexible frameworks.

The Future of a Curriculum and the Interpretive Campus

The most daunting curricular challenge is promoting understanding of how global environmental change (and its economic, political, and social impacts) will summon an unfolding array of unanticipated challenges. These will be both academic and conceptual (influencing how we think about sustainability) and applied and practical (influencing how we implement sustainability). That's why I propose an adaptive curriculum that emphasizes four flexible, yet substantive foundations—biosphere studies, social networking and change management, the creative imagination, and sustainability life skills.

As a response to rapid changes in the world of higher education, colleges and universities will increasingly encounter new approaches to program implementation. They will have to find the right mixture of visceral and virtual learning, residential and non-residency approaches, partnerships and sponsorships with businesses, governments and non-profits, consortia with other institutions, and international arrangements. These opportunities and challenges also raise difficult questions about academic independence, free inquiry, and curricular standardization. I suggest that institutions will have to maintain an adaptive approach here as well—prepared to innovate and diversify without succumbing to opportunism or mission creep. This is why it is important that campuses emphasize sustainability with all their constituencies, particularly governing boards, donors, alumni, and families. Colleges and universities should develop programmatic approaches and campus planning scenarios that integrate sustainability initiatives with the cultural tradition and academic strengths of the institution. Senior leadership, in consultation with staffers, faculty members, and other constituencies, must also assess what programs will meet the needs of their students and consider how to ensure the financial viability and the academic integrity of those programs.

Curricular design is relevant to multiple institutional decisions, from revenue to academic integrity. Ultimately it must be responsive to changing external circumstances. There is always public interest in curricular decisions. As a sustainability design studio, the campus has a public responsibility to explain what it is doing and why its sustainability initiatives are important. An interesting campus is replete with dynamic architecture, artwork, museums, public forums, landscapes, play spaces, squares, terraces, and quads, all contributing to a campus narrative, telling stories about student life, the cultural and ecological history of the place, and the values of the institution. They serve interpretive functions, conveying explicit and implicit ideas about life and learning. These public settings have a powerful effect on all campus visitors, and if properly designed serve as influential centers of learning.

8
Interpretation

All Visitors Are Students

Even in rural Maine, a college campus receives scores of visitors every day. There are admissions tours, parents and siblings, vendors, community members, gatherings and conferences, or people just passing through. I enjoyed meeting everybody because it gave me a chance to usher a warm welcome while briefly pointing out some of our sustainability initiatives. Whether I was speaking with a small group of prospective students and their families or hosting a gubernatorial primary debate, I cherished these meet and greet opportunities. Of course I got to meet only a small percentage of the many visitors. I wanted to make sure that everyone who visited campus was not only warmly greeted, but also had an informative learning experience.

Most people think of college and university campuses as places where a great deal of learning takes place. Despite all the scrutiny higher education receives, campuses have elevated stature in most communities. People expect a learning experience of some kind when they visit a campus, even if they are just passing through. And almost all campuses hold events that are specifically designed as learning experiences. Unity College was a center of learning in a small community and had an ambitious schedule of educational events. Larger campuses provide an exceptional variety and depth of such events—concerts, conferences, sporting events, public talks, public gardens, walking paths, museums, and so on. Some campuses have hundreds of thousands of visitors per year. Most of these visitors are expecting an enriching experience. They may not be matriculated students, but they are equally ready to learn.

At Unity College, I wanted every visitor to come away from campus with a better understanding of sustainability. We built that awareness into our campus tours. Most visitors would ask questions about the passive solar Unity House, the vegetable and wildflower gardens, and the wood chip boiler. I felt that unless we explained what we were doing and why, we weren't making sufficient use of our campus as a learning experience. Campuses everywhere have the opportunity to engage in comprehensive sustainability interpretation, and, in the process, to broaden visitors' awareness of ecology, natural history, and global environmental change.

Consider the experience of a visitor to a National Park. At the entrance gate he or she is given a map that highlights places to visit and tour. The park's interpretive center features educational displays explaining the ecological, historical, or geological setting that makes the place so special. What if college campuses were to take a similar approach? Upon arrival on the campus, a visitor would be given a map and a guide to all the campus's sustainability efforts. There would be tours, exhibits, and recommendations to visit certain buildings or distinctive sustainability projects. Signs would emphasize these sustainability initiatives, and kiosks would offer detailed and engaging information.

A campus is a geographical setting, with unique environmental and cultural features. If the campus is in the desert, how does the ecological setting determine the patterns of water use? If it is in a cold climate, how is innovative energy design used to help the campus stay warm while reducing carbon emissions? With imaginative signs, curriculum, website exhibits, and campus publications, this information may be fully incorporated in the campus tour.

In this chapter I'll describe how to use sustainability interpretation as a campus educational tool. First, I'll discuss the meaning of interpretation as both an educational methodology and a campus narrative. What is interpretation and why is it necessary? Who are the interpreters? Second, I'll present some principles of interpretive educational design that can become the basis for campus-wide sustainability initiatives. Third, I'll consider the campus in an ecological context, demonstrating how the sustainability ethos inspires discussions of global environmental change, biospheric awareness, and natural history. Fourth, I'll explain how the built environment provides additional learning opportunities and how we might conceive of buildings as teachers. Fifth, I'll present some guidelines for an idealized campus sustainability tour.

What Is Interpretation, and Why Is It Necessary?

In the field of environmental education, "interpretation" refers to a specialization that emphasizes how to maximize a learner's ability to better understand, observe, and make sense of the natural world. In phenomenology, "interpretation" refers to how people make meaning out of their everyday experiences.[1] Although these brief descriptions include many different perspectives, approaches, and intentions, what they have in common is their interest in the close relationship between narrative and meaning making. For the environmental interpreter every landscape tells a story. The task of the educator is to allow that story to evoke an ecological narrative in the mind of the learner. Though the narrative may be peculiar to the learner's experience, it may lead to a deeper understanding of the ecosystem. For the phenomenologist, the contingencies of direct experience and sensory impressions are the foundations for the making of meaning, the stories people construct about their lives, their daily narratives and worldviews. Many environmental interpreters, especially those who are attracted to experiential learning, find philosophical support in phenomenology.[2]

This chapter elaborates on the concept of interpretation as intrinsic to the sustainability curricular agenda. If the campus is perceived as a sustainability design studio, then interpretation should be an intrinsic feature of the campus narrative. Every campus strives to explain itself to its constituents and to visitors. It recounts its history, proudly trumpets its traditions, expounds its "unique" programs and accomplishments, and spends considerable resources to ensure that those stories are sufficiently and attractively promoted, branded, and displayed. In these ways, it is heavily engaged in methods of educational interpretation. Enrollment managers always desire the latest and greatest social-media techniques and printed materials so as to promote their campuses to prospective students. That's why institutional advancement offices invest their resources in sparkling videos and portfolios. In this chapter, we're taking a leap of imagination. What if the campus invested the same energy and attention on promoting its sustainability initiatives as the center of its interpretive narrative?

Just as advancement and enrollment offices urge a school's staff members and faculty members to embrace their work, and project their support of the institution, so I am suggesting that all members of the campus community focus on sustainability as an interpretive narrative. The campus itself can serve as a sustainability ambassador, providing visitors, residents, and employees with a hands-on educational experience. This requires that everyone associated with campus communication efforts understands the importance of sustainability initiatives and can explain what they mean and how they work, or at least how they affect daily life on the campus.

I recognize that this is an idealized view. A college or university campus is a complex place, embodying diverse narratives, populated by community members who may have different perspectives, manifested in a variety of stories. Indeed, students and alumni of different generations, with different cultural backgrounds, political perspectives, and values will interpret the campus according to their unique experiences. However, there are traditions, events, buildings, and places on a campus that unify those experiences, serving to integrate the campus narrative in colorful, symbolic, and engaging ways. The challenge for advocates of sustainability is to recognize the importance of those integrated narratives, understand why they are so important to the symbolic life of the campus, and find ways to utilize them so as to help advance the sustainability curricular agenda. New sustainability initiatives, whether they are buildings, retrofits, energy installations, community gardens, or restored habitats, can revitalize the campus's narrative, contribute to a new campus identity, link the past and the future, and change how people perceive the campus.

Interpretation is necessary because it represents the very essence of meaningful education, the awareness that learning can occur anywhere and anytime, not just in the classroom, but in the broader "field" of experience—in this case, the entire campus landscape. If there are solar panels on the roof of an administrative building, they represent an interpretive opportunity. Why are they there? How did they get there? How do they work? What do they signify? What choices do they represent? What stories do they tell?

Good interpretation simultaneously builds transparency and accountability. How do we measure the energy efficiency of the solar panels? How do we explain the concepts behind our data on energy efficiency? What is the best way to inform the interested public? How do we engage those who are less interested? How do these panels contribute to climate action planning on our campus? Why are we involved with climate action planning to begin with? A robust interpretive approach raises all these questions, none of which may be self-evident, many of which will prompt new questions, and even controversies. Isn't that what good education is all about?

An interpretive approach recognizes the necessity of incorporating diverse perspectives. The term itself connotes explanation, conceptualization, explication, and perception. There is not necessarily a "correct" interpretation. But there are really good methods that probe the meaning of experience, generate insight, and stimulate engaging conversations. These methods are the essence of evocative interpretation. A really good curriculum starts with good questions, offers a syllabus that allows the learner to probe those questions in engaging ways, and then equips the learner to ask a new set of good questions. How might campus-wide efforts utilize interpretive methods to broaden sustainability awareness?

Evocative Interpretation: Rules of Thumb

By "evocative interpretation" I mean educational approaches that elicit engagement, conversation, understanding, and a broadening of perspective. The challenge for sustainability interpreters is how to encourage those responses in informal settings. Ideally, a campus presents a seamless web of learning between the formal classroom and all aspects of campus life. However, many visitors to a campus are not matriculated students. In view of the diversity of the people who visit a campus, the various reasons why they come, and their range of backgrounds, figuring out how to summon their attention, hold it, and then construct a riveting learning experience is a tough challenge. How do you focus the attention of a casual visitor? Parks, museums, science centers, and public places face similar challenges.

Before I discuss my recommendations for the substantive content most worthy of interpretation, I'll lay out some general guidelines for engaging the casual visitor. Consider these guidelines for evocative interpretation. Although I consider these guidelines useful for a variety of subjects and situations, they are constructed with our subject in mind—the sustainability curricular agenda. This is a short list of engaging interpretive methodologies. I encourage their use in the spirit of hands-on, educational improvisation.

(1) Ask relevant questions. Learners are most likely to be engaged if you ask questions that are relevant to their experience, and if the thought process engendered by the questions enriches their experience. A skilled

educator understands that the same question is not necessarily relevant for everyone. It takes great skill, awareness, and finesse to develop questions that are of sufficient common interest, or to provide a sequence of questions so as to motivate diverse interests and learning styles. A good interpretive design will present questions that integrate all of those possibilities and also strive to generate interest from the otherwise disengaged. But before a question is relevant, you have to find a hook.

(2) Make the ordinary extraordinary. Routine experiences are the activities of everyday life, the things that you do all of the time—basic functions such as eating food and drinking water, moving from one place to another, heating and cooling a home or workplace, sitting in a chair, using a computer, and shopping. These basic functions are often interesting to people when they are challenged to consider them differently. How do you make those routine experiences more interesting? One way is to demonstrate the far-reaching circumstances of what we typically take for granted. The highly regarded "stuff" activities are a neat template for this approach.[3] By looking deeply at commonplace activities—tracing a drink of water from precipitation to tap, tracing the flow of energy when you turn on a light switch, investigating the full range of transactions when you buy a pair of running shoes—you gain an entirely new perspective. The depth and complexity of these activities is surprising, engaging, and illuminating.

(3) Call attention to scale. When you compare variables such as size, time and distance, you expand the boundaries of learning. This is a useful technique for showing the relationship between local and global issues and events. We tend to think in limited time frames, and restrict our awareness to what is directly in front of us. For example, if you consider the temporal dimensions of energy production, you realize that every energy source reflects a specific period of earth's history. Fossil fuels are the perfect example—ancient plants compressed over geological time power our cars and heat our homes. By following the biogeochemical pathway of a carbon atom, you cover a great expanse of biospheric space and time, and learn about climate change, biogeochemical cycles, photosynthesis, food production, and ultimately food and energy policy.[4]

(4) Elicit wonder and mystery. The most engaging learning experiences spark glimpses of wonder and mystery, a momentary awareness of the vastness of possibility. If that seems too grandiose, ponder the times when you have had such experiences. Chances are they accompany situations

when you've had a new insight about space and time, or you've witnessed something from an entirely new perspective. Many events or processes can trigger such experiences—from the grandeur of a sunrise to a riveting musical riff, or perhaps a moment of quiet contemplation, or a snowstorm that brings sublimity to a busy city street. Evocative interpretation should strive to facilitate such possibilities.

(5) Balance the visceral and the virtual. There are many splendid information technologies or other forms of instrumentation that bring a "wow" factor to interpretation. Technology has the capacity to instantaneously bridge space and time, thus enhancing an appreciation of scale, and to gather, assess, and manipulate data, lending a hands-on appeal to learning, facilitating simulations, models, and experiments. However, sustainability calls attention to the human relationship with the biosphere. Our direct encounters with the "natural world" are always close at hand, wherever we may be. Evocative interpretation finds ways to blend these direct encounters (visceral experience) with the use of technology (virtual experimentation).

(6) Highlight community practice. Living a sustainable life involves personal routines and behaviors as well as community practices. Campuses can emphasize all of the different ways that sustainability initiatives are enhanced, fulfilled, and catalyzed by community action. Evocative interpretation presents opportunities to cultivate community practice by explicitly demonstrating where these practices occur, and most specifically how individual actions can contribute to a greater good.

(7) Bring the story forward. The sustainability ethos calls attention to a compelling campus narrative, linking special places, traditions, and events to the daily practices of living and learning. Every place and building on campus has an interesting story, in many cases connected to a formative experience. What campus stories do these places tell? How do they form a larger narrative about the history and trajectory of the campus? How might they be woven into a dynamic sustainability narrative?

A Campus Is an Ecological Place

Every campus is an ecological place that embeds cultural meaning. Its geological features, its climate, and its flora and fauna are modified by human landscaping, and then fitted for residences, offices, classrooms, athletic fields, and all types of buildings designated for a variety of functions. Both outdoor and indoor places, by virtue of their natural beauty, human design, or the circumstances of their history, also take on deeper cultural meanings. One can learn a great deal about a campus just by observing where people gather, which places are most popular and which paths they use to move between destinations. The purpose of a campus master planning process is to maximize the efficient utilization of the campus for all of the functions cited above. With the emergence of the sustainability agenda, ecologically oriented planning processes are increasingly prominent. Still, few campuses call public attention to the ecological features of their natural and built landscapes. Why is it important to do so? And why is it crucial to interpreting sustainability?

In chapter 7, I claimed that the ability to perceive and interpret global environmental change is the curricular substrate for sustainability. This educational challenge requires an emphasis on ecology, evolution, and earth-system science. In a previous work (*Bringing the Biosphere Home*), I suggested that place-based natural-history observation is the most accessible, tangible, and effective way to learn about global environmental change.[5] Paying close attention to local flora and fauna, physiography, climate, and landscape is the conceptual key to broadening an understanding of spatial and temporal variation, juxtaposing scale, and recognizing biospheric patterns. Interpreting sustainability also starts here. Awareness of natural history and the biosphere, of humans and the ecosystem, and of ecology and evolution is the basis of any effective sustainability curriculum.

The challenge for sustainability interpretation is how to call attention to these concepts in all aspects of campus life, and how to do so given all of the diverse constituencies, perspectives, and agendas that pervade any campus community. In the remainder of this chapter, I'll suggest a catalog of approaches for cultivating an ecologically based understanding of sustainability, from buildings to the biosphere and then back again.

The singular contribution of the last fifty years of environmental education is the recognition that place-based learning is a highly effective way to motivate learning. From Gary Snyder's classic essay "The Place, the Region, and the Commons," to David Orr's original concept of ecological literacy, to the essays of David Sobel, Louise Chawla, and other educators who research childhood and place, to Keith Basso's anthropological investigations, ranging to all corners of the social sciences and humanities, we have accumulated compelling evidence suggesting that an orientation to place motivates learning for all subjects, not just ecology and natural history.[6] By recognizing the campus as an ecological place, we establish a conceptual starting point for sustainability interpretation.

Campuses are very dynamic places, offering an interesting blend of the permanent and the transient. For the student residents, the campus is a temporary home, and they come to know it intimately. For members of the staff and the faculty, the campus is a workplace, and they spend at least half of their day on its premises. For visitors, the campus is a destination—for business, culture, athletic events, or even tourism. Campuses typically have parks, museums, stadiums, libraries, concert halls, laboratories, shops, and hotels. Even the most private campus can be a very public place. People go out of their way to visit attractive campuses. The campus is a gathering point, a crossroads, and a meeting place. In many communities, the campus is the mark of stability, legacy, and relative permanence, with some of the oldest buildings in town. Yet its human residents are typically transient. And its buildings and landscapes may change as well, serving different functions and purposes depending on the priorities of campus space use.

All these dynamic functions are layered on an ecological and geographical place. The human residents are dwelling in a vital community of local and migratory flora and fauna, navigating the built and natural landscape through the changing weather and seasons. Astute observers will become familiar with these layers, deepening their appreciation of its richness and complexity.

These layers of buildings, habitats, organisms, and minds are the common ground for individual stories and experiences. Through those experiences, some places will take on special meaning—where people form or sever relationships, where they most productively study, where they have inspiration, or where they learn about love and loss. These encounters might represent seminal moments in a person's life experience, or they might serve as oblique memories in an ephemeral time. Either way, meaning unfolds. The task for the sustainability interpreter is how to find those meaningful places, amplify and elucidate their special quality, and then further enhance their ecological and cultural value—the unfolding layers of sustainability interpretation.

The Way of Natural History

Natural history is a great foundation for sustainability interpretation.[7] Who lives here? How long have they been here? Where did they come from? Who calls this place home? You start by knowing where you are, learning about your co-inhabitants, and better understanding how your place came to be. You observe the wildlife, the sky, and the landscape. This knowledge and awareness has been fundamental to human survival for most of human history. Almost everyone demonstrates some kind of interest in wildlife, weather, or the landscape, either as a casual observer, or as a way to interpret place, meaning, and identity.[8]

Natural history is the first layer of sustainability interpretation. It is the basis for understanding the human relationship with the biosphere. If you don't understand how your actions affect the environment, or how the biosphere is the source of all nourishment, energy, and habitation, then you won't understand sustainability. Natural history is the most accessible, tangible, place-based way to gain access to this awareness. Yet we are so often distracted from closely observing the natural setting.

Consider the campus as a natural-history field guide. The forms of communication may be kiosks, signs, displays, or even smart-phone applications that introduce the flora and fauna of the campus, including birds, mammals, insects, trees, and flowers. Possibilities may include pointing out wild edibles, distinguishing between permanent and temporary wildlife residents, illustrating bird migration routes, or depicting prey-predator relationships. Display landscape features by constructing maps featuring biogeographical patterns, geomorphological highlights, geological origins, and settlement patterns. How have these changed over the years? Remind people about the weather by illustrating prevailing winds, seasonal variations, likely cloud formations, and the relationship between weather, landscape, habitation, and living things. Has the campus been the site of an extraordinary natural-history event in the recent (human history) past—a catastrophic storm, a drought, or an invasive species.

Natural history is a conceptual bridge for illustrating the ecosystem services concept. Are there ways to show links between natural history and the ecosystem so as to illustrate the campus's eco-capital—its energy, water, food, waste, or biodiversity assets? How have those assets changed over time? How does basic knowledge of natural history enhance our awareness of those assets?

For a superb illustration of this approach to campus natural history, see Eric Sanderson's book *Manahatta: A Natural History of New York City*. Sanderson was asked by the Wildlife Conservation Society to reconstruct the ecological setting and history of Manhattan as the basis of a museum exhibit. He uses his expertise in landscape ecology to develop a portfolio of historical maps depicting the changing features of Manhattan, juxtaposing virtual images of pre-settlement Manhattan with the latter-day city. He constructs a variety of beautiful maps featuring the original topography, fill and excavation, original soils, streams, ponds, and springs, Lenape (native dwellers') sites and trails, human habitat suitability, locations of native dwellers' fires, and ecological communities. (Manhattan once had forty-five ecological communities within three-fourths of a mile of its shores.) Sanderson's final chapter projects how the Manhattan landscape will look in the year 2409. Sanderson's project is of great interest for any campus that wishes to illustrate its history as an ecological place. His book is filled with splendid and instructive interpretive templates.

Deep Space and Time Embedded in the Present

Sanderson's time frame is approximately 800 years, roughly from 1600 (the origins of European settlement) to 2400 (the equivalent distance into the future), specific to a narrow island. To fully grasp biospheric patterns and their relationship to local place, it is necessary to interpret broader vistas of space and time, extending to the geological time scale and considering the geography of the whole earth. Embedding biosphere-scale space and time in the present is a difficult, important, and highly engaging interpretive challenge. Bringing the biosphere to the campus is the second layer of sustainability interpretation.

What does this mean? How do you link biosphere-scale phenomena to local natural history? This is an educational challenge of great urgency and should be on the agenda of all educators who are concerned about the future of the planet. I have written about this challenge at great length elsewhere.[9] For now, we can limit the question to campus sustainability efforts. For example, one of the biggest obstacles to understanding climate change is that few people fully grasp the depth and splendor of the geological time scale. When a climate scientist offers scenarios that suggest unprecedented century-scale temperature increases, the magnitude and severity of the changes are lost on people who have no scalar context for this understanding. Practically speaking, most people, most of the time, think in extremely limited time frames. An intergenerational perspective is hard enough. A geological-time-scale perspective is just not intrinsic to everyday awareness.[10]

Where does all of this fit in the campus sustainability narrative? Biosphere-scale conceptualizations can be incorporated in routine interpretive approaches. Natural-history observations and displays can involve evolutionary and geological concepts. Discussions of the built environment (see the next section), whether they entail interactive exhibits regarding water, food, waste, or energy, can easily provide a broadened time and space perspective. The geological time scale can be permanently displayed in prominent settings.

There are precedents for these approaches in museums and planetariums. I am suggesting that these conceptualizations should be more frequently embedded in the routine behaviors of everyday life. To further imagine the possibilities, I invite you to find your favorite search engine, first type in "geological time scale," and follow that with "tree of life" and then "biosphere." You will find thousands of compelling, downright inspiring illustrations, concepts, and visualizations. Such illustrations can be a foundation for campus sustainability interpretation. Other approaches might include exhibits of relic environments, extinct species, or different geological eras and ecological regimes. With imaginative art and sound science, and using the principles of evocative interpretation, these are pathways to grasping scale, enhancing the campus sustainability narrative, and deepening the possibilities for interpreting global change.

Interpreting Global Change

The most effective way to "bring the biosphere to campus" is to demonstrate how the routines of everyday campus life can be linked to biosphere-scale phenomena. For example, campus monitors can trace phenological patterns (the coming and going of the seasons), noting the arrival and departure of species, the blossoming and fruiting of plants and fungi, insect blooms and their relationship to species migration, the growing season of gardens, and how daily weather reflects global and regional climate patterns, including the monitoring of wildfires, drought, floods, or other unusual events. Observing global environmental change is the third layer of sustainability interpretation. How might the campus promote such awareness? Here are a few ideas for getting started:

Set up interactive campus monitoring stations that display ongoing environmental change. To the extent possible, get those monitors out of the lab and onto public campus places. Whether with simple (and inexpensive) signs that compares, say, phenological events through a geographical sequence, or with more detailed exhibits linked to global research networks, make these observations public. Highlight the work of faculty (and student) researchers who are investigating global environmental change.

Distribute interactive computer applications (for example, inaturalist[11]) or other computer resources having to do with global environmental change (for example, Google Earth[12]) throughout the campus. Provide instruction in their use at orientation sessions or at public kiosks. Explain how they can be linked to campus-wide monitoring of global environmental change. Promote campus-wide events to build awareness of and pride in the campus as an ecological place. For example, you can organize annual or semi-annual "bioblitzes," supervised by members of the natural-sciences faculty. A bioblitz is an intense 24-hour survey of all living things in a designated area. The objective is to increase public interest about biodiversity. There is a global movement of bioblitzes.[13] The campus is an ideal environment for such events, and as the basis for global environmental change monitoring and record keeping. Develop a campus catalog of species, and an accompanying tree of life.[14]

Use campus publications, maps, and websites to illustrate the environmental-change "news of the day." Develop easy-to-read charts (like a simple weather forecast) that appear daily in the campus newspaper, or on the front page of the campus website. Provide this information in annual reports, alumni bulletins, and other widely distributed publications. Encourage the laboratories, museums, and art studios to invite visitors from all corners of campus to participate in, observe, and learn from their sustainability and environment-related research. Challenge these facilities to organize environmental change experiments, demonstrations, and exhibits.

Consider how buildings can become sustainability monitoring stations, recording, and demonstrating how they use energy, materials, and water. How might the variety of campus buildings serve as ways to remind their inhabitants that every breath is an exchange with the biosphere, that every material action is a biogeochemical exchange, and that water use and energy represent a complex cycle of physical, chemical, and geological processes?

Buildings as Extensions of Human Awareness

I have tried to develop a case for why a better awareness of local natural history is the foundation for understanding global environmental change, and how this might serve as a template for interpreting sustainability. This awareness is built on the importance of "outside" learning. Yet no matter how much we might advocate that people should spend more time outdoors,[15] the fact is that most people, especially college students and campus employees, spend the great majority of their time indoors. For the most part they dwell, work, study, learn, and recreate indoors. With the exception of windows (real ones, not the computer software of the same name), most indoor environments create a stark separation between indoor and outdoor spaces.

Yet buildings are constructed from "natural" materials that are heated, treated, manipulated, refined, and shaped for human habitation. From bird's nests to skyscrapers, sheltered spaces are ways of gathering the materials of nature for various adaptive purposes. However, it is much easier to follow the construction path of a bird's nest than to understand all the industrial processes, infrastructure complexities, and architectural intricacies of a skyscraper. If you live in the equivalent of a bird's nest, let's say a wooden tree house, you have a very clear idea of how your shelter was made, and you experience a seamless relationship between inside and outside.

Bringing the outside inside is the fourth layer of sustainability interpretation. There is a rich tradition in ecological architecture, in city and regional planning, and in sustainable design that emphasizes the utility, efficiency, and aesthetic of this approach, culminating in the "living building concept."[16] The U.S. Green Building Council defines a living building as a structure that "generates all of its own energy with renewable nontoxic resources, captures and treats all of its water, and operates efficiently and for maximum beauty."[17]

Type "living building" into your browser, ask for images, and you'll be treated to a remarkable portfolio of ingenious designs, adapted to a variety of functions, habitats, and environments. Some of these are on college and university campuses. A good example is Oberlin College's Adam Joseph Lewis Center for Environmental Studies. In conceiving this building, Professor David Orr and his colleagues understood the necessity of using the building for educational purposes.[18] Hence visitors and occupants alike are always reminded of how the building is an interface between humans and the environment.[19]

At Unity College, as I have already noted, we built a zero-carbon president's residence. Unity House was designed to transparently demonstrate daily energy use and to show how recycled materials were embedded in its construction. We knew that the president of a college hosts numerous events and gatherings. The house was designed to provoke questions about what it means to live a sustainable life. As the president, I was aware that I had an unprecedented opportunity to engage in sustainability interpretation. We spent hours giving tours of the house to people who were interested in sustainable construction. As we actually lived in the house, people were fascinated with the interface between the building and our daily lives. We also constructed student housing using the Terahaus system. Our long-term design philosophy at Unity College was to turn every building into an educational experience.[20]

Buildings are extensions of human awareness. When they are designed in harmony with the landscape, when they follow principles of ecological design, when they use materials that are appropriate to the setting, when they circulate clean and fresh air, and when they are filled with penetrating light reflecting the changing rhythms of the day, they take on a "life" of their own.

Buildings as Teachers

Using buildings as teachers is the fifth layer of sustainability interpretation. How is this accomplished?

First, any building can be a teacher. It doesn't have to be a sparkling new ecological design. Most campuses are filled with older buildings or energy-inefficient buildings that were constructed during times of cheap and abundant energy. There is educational value in explicitly pointing out what is wrong with these building designs and how a thoughtful retrofit processes can maximize energy efficiency.

In the late 1960s, when oil was cheap, Unity College had a multipurpose campus center built. It was so poorly insulated that some energy companies were hesitant to take Unity on as a client. We used to joke that it heated the outdoors—no small feat in rural Maine. It was our most egregious emitter of greenhouse gases. In one respect, this building was one of our best "educators": it was a daily testimony to poor design. One class developed an exhibit called "a museum of inefficiency." Since we didn't have the funds to replace the building, we spent hours thinking about how to make it more energy efficient. I cite this example because many other colleges and universities also have grossly inefficient buildings. It is great fun to explore new designs, but what do you do with the old buildings? I am suggesting that sustainability interpreters would do well to pose this challenge as publicly as they can.

Second, buildings can use monitoring technology and make the data readily available. For example, all campus buildings are typically assessed in a campus's greenhouse-gas-emissions plan. Typically, those assessments are just figures on a spreadsheet. Make them public by developing displays of that data in those buildings. Provide public monitoring stations that allow residents and occupants to trace and chart the energy use of every building. Extend the same approach to other aspects of the building, including air circulation and freshness, or access to light. It is also helpful to explore the ways buildings are actually used. Where are the public spaces? Where do people congregate? What is the relationship between building use and energy use?

Third, consider the use of materials. Buildings use water, minerals, metals, plastics, glass, and energy. Where do those materials come from? How much energy is used to produce them? What waste products (CO_2, for example) result from their use? Identify, analyze, and interpret all of the materials used in the construction of the building. What materials are recycled? Where do they go? How are they reused?

Fourth, all buildings are filled with living organisms. What other living things inhabit the building—for example, small mammals, birds, insects, and microbial organisms? Where do they live? How do they enter and leave? How do they affect everyday life in the building? How do they respond to air, light, and heat circulation? What do they eat? What plants live in the building? Are they native? If not, where do they come from? How did they get here? What is their evolutionary history? How do they improve the quality of life?

Ecological architecture emphasizes the "living building" concept as an aspiration for sustainable design.[21] It lays out a series of design criteria for creating a seamless relationship between a building and the biosphere, from the most basic indoor/outdoor interface to using a building to "inhale" carbon emissions. William McDonough, in an article titled "Buildings Like Trees, Cities Like Forests," presents a compelling vision:

[W]hat if buildings were alive? What if our homes and workplaces were like trees, living organisms participating productively in their surroundings? Imagine a building enmeshed in the landscape that harvests the energy of the sun, sequesters carbon, and makes oxygen. Imagine on-site wetlands and botanical gardens recovering nutrients from circulating water. Fresh air, flowering plants, and daylight everywhere. Beauty and comfort for every inhabitant. A roof covered in soil and sedum to absorb the falling rain. Birds nesting and feeding in the building's verdant footprint. In short, a life-support system in harmony with energy flows, human souls, and other living things.[22]

This is a wonderful vision, but very few campuses will have the resources to pursue it. However, a campus can prepare the way for "living building" concepts by showing how every building has such interfaces, whether they are "harmonious" or not. The challenge for sustainability interpretation is to demonstrate all the ways in which buildings themselves are a manifestation of biospheric processes— materials exchange, energy flows, water movement, and life support processes. This facilitates a deeper interaction with a building, hopefully fostering ecosystem-based improvements in the quality of life.

Campus Sustainability Tours

Since 2010, dozens of colleges and universities have been offering campus sustainability tours. Because this is a new trend, I expect that you can best keep current on it by periodically typing "campus sustainability tour (or map)" into your browser. As of this writing, a survey of these maps and tours reveals a wide range of approaches at a wide variety of colleges and universities.

Of great interest is how tours are used to make the point that sustainability is deeply rooted in the mission and values of the campus. For example, Colorado College claims that it aspires to "nurture a sense of place and an ethic of environmental sustainability," and that "sustainability isn't optional for the Colorado College Community; it's who we are and how we have defined ourselves."[23] Portland State University has an outstanding walking tour that emphasizes the alignment of mission, landscape, and values:

At Portland State, we strive to have everything we build teach us, and everything we learn help us create a better world. Our 50-acre urban campus is a vibrant, living laboratory for practicing sustainability, extending outside the classroom into offices, eateries, plazas, and gardens. In the coming decades, our goal is to be a model of sustainability not only for a university campus, but also for the surrounding neighborhood, city, region, and even the globe.[24]

At the University of California at Davis, a campus sustainability map was initiated as a way to comprehensively integrate a variety of initiatives. "The interactive map can be used to find places or things on campus that express or demonstrate ways that UC Davis and the campus community are taking action towards creating a more sustainable future." The map is conceived as a living document, soliciting contributions from student teams, classes, and all campus constituents. "Future goals for the sustainability map include using the map as a springboard for creating tours, adding downloadable maps, and adding more points and more information about campus infrastructure and sustainability."[25]

A campus sustainability tour (and map) should be central to any interpretive strategy. It is a way to focus attention on the whole campus system, integrating the full range of initiatives, linking these efforts to a coherent educational philosophy, serving a variety of instructional purposes. Such tours are ideal venues for community participation, visitor orientation, and public outreach. They also send a solutions-oriented message to all campus constituents. They build the concept of sustainability as a value proposition—the campus as a cradle of sustainability innovation or a sustainability research center.

I hope that businesses, health-care facilities, museums, and places of worship—any institution that has a campus—will initiate similar tour processes, and that as a result comparative interpretive methods will become increasingly important and interesting. But how good will the sustainability tours be? Will they be sufficiently evocative, informative, and interactive? Will they link sustainability to concepts of ecology and the biosphere? Will they seamlessly link the everyday life of the campus to the broader vision and values of sustainability? Will they juxtapose social media, interactive computer applications, and hands-on sensory learning? Will they sufficiently utilize artistic approaches? These challenges will be faced by a new generation of sustainability interpreters as they unleash their skills and experiences in a multitude of campus environments.

9

Aesthetics

The Art of Stewardship

In the autumn of 2008, Unity College invited sixty artists, sustainability practitioners, students, humanists, and scientists to a small conference. It was a cool, rainy, windswept November day. The participants were crammed into the public living space of Unity House, the LEED platinum president's residence. We opened the program with a challenge: How can we turn Unity College into a campus canvas for art that conveys, expresses, and inspires ideas about sustainability, stewardship, and ecology? After some brief opening comments, we immediately engaged the participants in a series of workshops. First we encouraged them to develop a portfolio of possibilities—a catalog of art installations for the Unity College campus. Then we asked them to spend the afternoon implementing some of their ideas. Small groups fanned out (in the pouring rain) across the campus, painting murals, assembling sculptures, or sketching plans for further elaboration.

The conference was inspired by the artist Greg Mort. Through the Art of Stewardship Project, he "brings artists, environmental and educational organizations together to use the power of imagery to build awareness about the earth's fragile ecosystems."[1] With support from the K2 Family Foundation, Greg and Nadine Mort, individual donors, and the guidance of regional artists, we aspired to transform the campus of the college into a canvas for interesting sustainability art. Our concept was simple: to inspire the campus community, especially students, to see the campus as an emerging work of art, a place where they can use creative expression to explore their ideas, concerns, and insights about sustainability.

For Unity College, the conference served multiple purposes. It promoted the college's visibility as a regional leader in sustainability and the arts, bringing new constituents to the campus, and encouraging an unusual network of people to work together. In so doing, it sparked many interesting new partnerships between the college and the community. It empowered students and faculty members to "play" with the campus, to conceive of the landscape and buildings as one big studio for sustainability art. It gave license to the prospects for public art on campus. It provided a venue for scientists, artists, and sustainability practitioners to work together. The conference inspired interest in arts-based curricular innovation, providing the faculty with ideas and support. As the president of the college, it gave me an interesting platform for carving an idea niche, for calling attention to what we do well, leading to good publicity and potential philanthropy. Finally, it made the campus much more physically attractive, and we were able to achieve a noticeable aesthetic upgrade very inexpensively.

When the conference was over, I held a debriefing meeting with members of the arts and humanities faculty. They were very enthusiastic about the day. We discussed what we could do to expand the effort, or to build it into the culture of the campus. I told them that they had free reign to engage their students and colleagues (both on campus and in the regional arts community) to construct any campus art installations that they thought would be of interest. This could involve student projects, art exhibits, performances, or other unique expressions. I emphasized that I wanted to be surprised. My expectation was that every few weeks I would see some interesting new public art project. In the months to come my expectations were fulfilled. Sculptures, murals, exhibits, and some more unusual works appeared. In just a few months, the campus was transformed from a rather pedestrian place to a very compelling setting. The ideas caught on. Students proposed wildflower gardens and other landscaping ideas. Other classes would use campus art projects to portray controversial issues. At very, very low financial cost, Unity was transformed into a much more interesting-looking campus.

The Art of Stewardship Project served many purposes, ranging from community engagement to enhancing philanthropic prospects. Yet the most important motivation was educational. The project's organizers had a deep-seated belief in using public art to express ideas, gather insight, stimulate the imagination, and deepen awareness about sustainability.

They were guided by an educational assumption that the artistic process enhances perception of the biosphere. There is a close connection between cultivating an aesthetic sensibility and observing the natural world. In this chapter, I'll explore that connection and explain its relevance to sustainability.

I'll start by suggesting that art enhances biospheric perception. Why is cultivating the imagination so critical for understanding global environmental change? Why is it important to create a bridge between art and science, and why is this central to the sustainability ethos? Art provides a setting for communicating emotional responses to what can otherwise be overwhelming issues. This is a foundation for creative sustainability. I'll investigate how art projects can inspire a campus and why public art has the potential to engage the campus community. I'll provide some suggestions for specific projects that incorporate the arts and sustainability. I'll conclude with a discussion of the sustainability aesthetic.

Art Enhances Biospheric Perception

Over the years, I have collected books about the biosphere, biogeochemical cycles, the history of life on earth, cloud formations, weather patterns, biodiversity, evolution, and natural history. I'm especially attracted to volumes that integrate lyrical text with interesting illustrations and photographs. I skim and study these works for multiple purposes. Sometimes I require accessible information about a topic of interest. Sometimes I need help understanding a concept. And sometimes I'm looking for inspiration. These books are extraordinary teaching aides. After spending some time in this "biosphere library" my imagination is nourished.[2] So inspired, my sensory awareness is heightened, and my readiness to "perceive" the biosphere is enhanced. I'm more likely to pay attention to phenomena that I typically take for granted. I have a broadened view of ecological space and geological time. I'm further compelled to use my artistic imagination to interpret, express, and communicate these impressions. I am in a state of enhanced wonder.

Most of the scientific concepts in these books are complex and difficult. To conceptualize theories of environmental change, evolutionary ecology, biogeochemical cycles, and the very concept of the biosphere requires sophisticated scientific training. How do you gather the evidence that allows you to locate a fossil in geological time, determine the movement of tectonic plates, figure out the location of ancient seas, trace the path of a carbon atom, or analyze the biochemistry of photosynthesis? Unless you devote your life to understanding these concepts, it isn't likely that you will grasp them fully. But that doesn't mean you can't develop a basic working knowledge. Nor does it rule out awareness of their depth, magnitude, and importance, and ultimately an aesthetic appreciation.

I admire the scientists who have the remarkable ability to help me understand these challenging concepts, those writers who have a knack for making complicated ideas accessible, interesting, and inspiring. How do they do this? The best science writers know how to immerse us in the wonder of their subject. They challenge us to visit other places and times, landscapes of imagination, enveloped in a narrative that brings us directly in contact with a place that we never otherwise experience. They accomplish this through lyrical narratives, engaging illustrations, and revealing photographs. Often they work closely with visual artists and photographers who are adept at conceptual illustration. They use metaphor, imagination, and scale. A brilliant photograph of the earth from space conveys extraordinary layers of information. A sequence of these photographs, with suitable annotation, enhances these layers. Portraits of Earth from space remind me of beautiful marbles I had as a child. Watching a sequence of swirling cloud patterns making their way across the planet reminds me of shadows traversing the landscape on a partly cloudy day. There are common forms, patterns, and images that stimulate the imagination, opening the perceptual pathway for deeper understanding of scientific concepts.

Let me explain the conceptual relationship between my biosphere library and the nine elements of a sustainable campus, and, in so doing, reiterate a prevailing theme. Sustainability initiatives are a response to concerns that human activity is presenting biosphere-scale challenges to human flourishing and biodiversity. The sustainability ethos promotes life practices that coordinate human activity with ecological and biospheric considerations. A basic understanding of biosphere processes is fundamental to well-conceived sustainability initiatives. Yet, as I pointed out in chapter 8, this understanding is conceptually challenging because it requires a broadened conception of spatial and temporal variation. How can sustainability educators bring this awareness to campus? How might this broadened awareness penetrate the day-to-day reality of campus life?

Not everyone has a biosphere library to offer ready inspiration. Nor will everyone choose to be inspired in that way. And it isn't necessary. We are the biosphere (in a human body). Our breath instantly connects us to biogeochemical cycles. A quick look at the sky can be a link to weather systems that cover an entire biome. The warbler in the oak tree may live in two hemispheres. Its semi-annual migratory journey is a reminder that its winter home is very far away.

Compelling artwork can transform everyday observations into moments of wonder and contemplation. The same illustrations in those neat biosphere books can be translated into campus art projects. Or other approaches can be explored—for example, soundscapes, dance, poetry, and electronic art. By utilizing the arts, the campus cultivates the imagination. But why is this a good thing?

Why Imagination?

The most extraordinary learning moments occur when we transcend the boundaries of expectation. We have a conceptual breakthrough that allows us to gather fresh insights and deepened awareness, stretching our capacities and capabilities. There is no formula for how this occurs. The challenge for the skilled educator is to construct learning environments that stimulate conceptual breakthroughs. There is a vast literature that addresses creativity, imagination, and innovation as relevant for such inspiration.[3] I will briefly describe three approaches that are of particular relevance because they call attention, in their own way, to the natural history of imagination. Then I will explore how this yields a foundation for creative sustainability.

Steven Johnson, in *Where Good Ideas Come From: The Natural History of Innovation*, is interested in what he describes as "the spaces of innovation" and how "some environments squelch new ideas" and "some environments seem to breed them effortlessly." His argument is that "a series of shared properties and patterns recur again and again in unusually fertile environments."[4] What is unique about Johnson's approach is how he links those patterns to multiple scales across the biosphere:

We have no shortage of theories to instruct us how to make our organizations more creative, or explain why tropical rain forests engineer so much molecular diversity. What we lack is a unified theory that describes the common attributes shared by all those innovation systems. Why is a coral reef such an engine of biological innovation? Why do cities have such an extensive history of idea creation? Why was Darwin able to hit upon a theory that so many brilliant contemporaries of his missed? No doubt there are partial answers to these questions that are unique to each situation, and each scale: the ecological history of the reef; the sociology of urban life; the intellectual biography of a scientist.[5]

Johnson suggests a "long zoom" vantage point, "a kind of hourglass" that ranges from nature (global evolution, ecosystems, species, brains, cells) to culture (ideas, workspaces, organizations, settlements, information networks). He concludes that "unusually generative environments display similar patterns of creativity at multiple scales simultaneously," and he advocates the importance of openness and connectivity, suggesting that "we can think more creatively if we open our minds to the many connected environments that make creativity possible."[6] He presents seven patterns that promote multi-scalar spaces of innovation, derived from a convergence of properties in both organizational and ecological settings.[7]

In *The Age of Wonder*, the historian Richard Holmes writes evocatively about the era of romantic science. He describes in detail how the process of scientific discovery combined the rigor of analytical experimentation with the insights derived from the creative imagination. In the era of romantic science, prior to disciplinary specialization, natural philosophy was a platform for the open exchange and connectivity of ideas. Each new discovery opened a world of wonder and amazement that inspired intellectual correspondences between great poets and pioneering scientists. When the astronomer William Herschel perfected telescopes that enabled him to penetrate vast vistas of stars and celestial phenomena, he captured the attention of William Wordsworth. When Humphrey Davy experimented with nitrous oxide to simultaneously open new approaches to chemistry and mind science, his work inspired John Keats. And then Mary Shelley (in *Frankenstein*) worried about what great powers had been unleashed. The milieu of romantic science hinged on the prospect of wonder. It was an era of astounding creativity because the spaces of innovation (to use Steven Johnson's idea) were multi-scalar, connected, and open-ended. Holmes quotes the brilliant Humphrey Davy, who wrote the following in 1807:

The perception of truth is almost as simple a feeling as the perception of beauty; and the genius of Newton, of Shakespeare, of Michael Angelo, and of Handel, are not very remote in character from each other. Imagination, as well as the reason, is necessary to perfection in the philosophic mind. A rapidity of combination, a power of perceiving analogies, and of comparing them by facts, is the creative source of discovery. Discrimination and delicacy of sensation, so important in physical research, are other words for taste; and love of nature is the same passion, as the love of the magnificent, the sublime and the beautiful.[8]

Holmes suggests that the era of romantic science was a template for cultivating a "wider, more generous, more imaginative perspective" and an urgent educational necessity. "Above all, perhaps, we need the three things that a scientific culture can sustain: the sense of individual wonder, the power of hope, and the vivid but questing belief in a future for the globe."[9]

In *Free Play*, the musician Stephen Nachmanovitch addresses the relationship between imagination and improvisation. A central tenet of "free play" is that the biosphere is a dynamic, unfolding array of emerging, multi-scalar patterns, a field of improvisation that engenders a personal response, and that the creative process reflects our ability to "play" in that "field" by exploring, experimenting, and connecting our individual interpretation to this field of experience:

Our body-mind is a highly organized and structured affair, interconnected as only a natural organism can be that has evolved over hundreds of millions of years. An improviser does not operate from a formless vacuum, but from three billion years of organic evolution; all that we were is encoded somewhere in us. Beyond that vast history we have even more to draw upon; the dialogue with the Self—a dialogue not only with the past but with the future, the environment, and the divine within us. As our playing, writing, speaking, drawing, or dancing unfolds, the inner, unconscious logic of our being begins to show through and mold the material. This rich, deep patterning is the original nature that impresses itself like a seal upon everything we do or are.[10]

These interesting and convergent approaches to imagination provide a conceptual foundation for an aesthetic and educational approach to sustainability. Why does this matter? The sustainability ethos presumes that a deep awareness of the human relationship to the biosphere will enable us to construct a more fulfilling, life-generating, creatively oriented community. Its premise is that adhering to the principles of sustainability will improve the quality of human life. Yet this will only occur if we can project a vision of what is possible, a landscape of new ideas and arrangements, and a campus that embodies these prospects. Sustainability conjures the art of the possible, and such a conjecture requires imagination. This is the essence of creative sustainability.

What Is Creative Sustainability?

In my early days at Unity, I was concerned that the college wasn't well known beyond the state of Maine, and that it lacked a distinctive academic signature. We had to find ways to increase our visibility. Many other colleges and universities were developing sustainability programs. How could we build on our strengths as a college, cultivate a unique academic niche, and simultaneously enhance our philanthropic prospects? I held several meetings with the humanities faculty. We decided that there was a need for a journal that could explore the artistic, philosophical, and poetic challenges inherent in living a sustainable life. We thought that such a journal would be an ideal venue for many writers and artists who cared deeply about sustainability issues. We discussed what kind of financial support it would need and how often it could be published. The faculty agreed that an annual budget of $10,000 would be enough to publish one issue a year. We negotiated the appropriate course releases, I allocated money from the President's Office, we organized a national editorial board, and we were ready to go. Thus was *Hawk & Handsaw* born. One faculty member (Kate Miles) was named the literary editor; another (Ben Potter) was named the arts editor. They were delighted to have their own journal. My only stipulation (from the President's Office) was that I would develop a mailing list that would include prominent writers and artists, leaders in the environmental and sustainability fields, and potential donors. From my perspective, not only was the journal a great way to stimulate ideas about sustainability and the arts; it also gave me a platform for calling attention to the good work coming from the college.

After several issues, we had gathered dozens of readers' responses to the question "What is creative sustainability?"[11] and we began to publish the best of them. In this section, I'm going to provide excerpts from some of the most interesting responses, and then synthesize their collective voice, elaborating and amplifying the potential for creative sustainability.

From Alan Crichton, a co-founder of Waterfall Arts in Belfast, Maine:

Creativity is the capacity for hard work, imagination and openness to new ideas, the capacity to see fresh connections between chance phenomena, to being open to accident and mistake and the unexpected as a means to new breakthroughs and solutions. The capacity for simplicity and elegance, for improvement, efficiency and beauty, for invention and enthusiasm. Creativity brings people together and encourages dialog, recognizes contribution and celebrates innovation. Creativity lightens the spirits. Creative Sustainability would be applying the fresh reach of creativity to the development of valuable new forms which change conditions for the better, make the world smarter, and bring clarity and happiness.

From John Tallmadge, author of *The Cincinnati Arch: Learning from Nature in the City*:

Sustainability is not a problem, a condition, or a program; it's a way of life, a relationship in which humanity and the rest of nature become, in the words of Thomas Berry, "mutually enhancing presences to each other." In this respect, sustainability resembles love, health, or peace. Pursued with deliberate imagination, it becomes a life practice for both individuals and communities. Think of sustainability as a type of infinite game, in which the goal is not to win (which would end the game), but to keep on playing forever. In practical terms, sustainability must always manifest itself in some place with some people; it always has a local, personal flavor. And because conditions and people change, sustainability always appears dynamic and evolving. It involves learning and transformation: this is where creativity comes in. You can't have sustainability without imagination.

From Sheryl St. Germain, author of *Let It Be a Dark Roux: New and Selected Poems*:

Creative sustainability marries imagination and political thought or action. It's an expression that acknowledges our need of the imagination's muscle to tell the most seductive stories, create the most evocative poems, and construct the most winning arguments in support of the earth and all of its denizens.

From Kurt Caswell, author of *In the Sun's House*:

The practice of making art in such a way that it may endure for a lifetime, and even beyond, if passed on to a new generation. It is not the art itself that is sustainable and so passed on, but the making of it. By art I mean not only writing, painting and dance, but also beekeeping, gardening, cooking, "going green," even conversation, anything at all that may be practiced and refined and so, highly performed. By such practice, one's life, rather than one's art, soon becomes the expression of creative sustainability. As the Buddhists are wont to say, the doing matters more than the reward.

From Amy Irvine, author of *Trespass: Living at the Edge of the Promised Land*:

It begins like this: We begin to imagine how our emotional and physical lives might defy every law of diminishing returns. Then we yearn: We grow ravenous with the desire not to consume, but to be consumed, by the human and biotic communities that support us. And finally grace: We stand in the sweetness of self-expenditure so that every other living thing bursts with ecological possibility.

From John Calderazzo, a professor of English at Colorado State University:

CREATIVE SUSTAINABILITY?

Aurora borealis scooped from

The sky, then locked in glass bulbs

To make our bedrooms glow in

Diaphanous green ribbons of light?

Pterodactyls regrown from DNA,

Coaxed by flocks of sandhill cranes

And white pelicans to drag sun shields

Through the burning greenhouse air?

A trained pit bull in every back yard,

Sweetly crunching bottle-mountains,

Cans, plastic jugs, word-weary laptops,

Metaphors of want and want and want?

A Yucca Flats for carbon capture, a

Sing Sing for oily appetites, jail time for

Hummers, river thiefs, ax murderers of

Old growth and the only home we have?

Creative sustainability integrates human and ecological possibility. The respondents aspire to derive life practices from their close observations of the natural world. The very process of doing so challenges their creative sensibilities. The biosphere is the field of play—the infinite variation of evolutionary ecology and earth systems, the recycling of life forms and biogeochemical cycles, and the perennial renewal of sustenance. These processes reveal unfolding patterns that stretch the potential of human creativity.

Creative sustainability involves the pursuit of virtue. It aspires to apply ecological principles and awareness to human behaviors and decisions, linking the quality of human life to the evolving biosphere. There is a prevailing assumption that by closely observing natural systems, we better understand how to live our daily lives, how to make a living, and how to coordinate our aspirations with the common good.

Creative sustainability is an emotional challenge. The intention of *Hawk & Handsaw* is to provide a voice for the dilemmas, ironies, contradictions, and inconsistencies that people encounter in pursuing a sustainable life. These challenges evoke interesting personal narratives, best expressed through artistic representations, providing the freedom to explore the depth of emotional responses—from elation to despair, from certainty to confusion, from purity to pollution, from the beautiful to the grotesque, from pleasure to abstinence, from acceptance to denial, from the extraordinary to the routine.

Creative sustainability requires community collaboration. Although the creative process is intensely personal, its public expression epitomizes community outreach. Pursuing a sustainable life, whether it is community gardening, installing a solar panel, retrofitting low-income housing, or riding a bicycle to work, involves people working together to improve community life. It requires honesty, clarity, and deliberation about common successes and failures.

Creative sustainability is an intergenerational process. It invokes long-term decisions about equity and distributive wealth, coordinated with resource extraction and ecosystem health. These deliberations extend in space (across the globe) and time (from the past to the future). Such bold and broad visionary thinking is enhanced by artistic narratives told from multiple cultural and generational perspectives.

If creative sustainability is the process, then what is the project? At Unity College, the Art of Stewardship Project helped us set an agenda for using art as a vehicle for campus transformation. *Hawk & Handsaw* was our way to express a collective vision for moving that process forward. In other campus settings, the project may take different forms.

The Campus Canvas

Sustainability art has the potential to creatively transform the culture of a campus. It can do this by tangibly illustrating sustainability principles in multiple settings, using a variety of artistic mediums, and engaging all campus constituencies in the process of making public art.

On many campuses, art displays are limited to designated spaces. "Accomplished" artists display their works in galleries, museums, and other dedicated facilities. Students learn and practice in art studios. Art becomes a specialized process. Yet just about everyone takes pride and pleasure in decoration. Students hang posters and photographs in their residence halls. Faculty and staff members often distinguish their offices with their choices of art and decoration. Making art is typically an individual matter as well. You do it as a means of personal expression, on your own time, as a form of enrichment. Art is perceived as extracurricular. Nothing typifies this more than the willingness of institutions to drop art programs at the first sign of budgetary stress. This is true in college and university settings as well as in K–12 school programs.

When the entire campus is a canvas, art projects are no longer confined to specified locations. Art projects might show up anywhere. They may be planned or spontaneous, permanent or ephemeral, structured or improvisational, professional or amateur, individual or collective, public or private, prominent or subtle. When these projects emphasize sustainability concepts, they will address place and landscape, natural history and the biosphere, conservation and efficiency, resource extraction and consumption, and hopefully promote behaviors influenced by the sustainability ethos.[12]

This is a challenging approach for many campuses, especially those that take great pride in their historic buildings, rely on experts to provide architectural and planning designs, and relegate art to designated spaces. However, for some campuses, an open approach to public art is a wonderful and cost effective way to promote creativity, vitality, and intellectual interest, while simultaneously developing an interesting campus aesthetic. Readers will have to assess the readiness of campus leadership to move forward on such projects. Other sections of this book provide leadership suggestions for how to mobilize such readiness.

The campus canvas, if planned well, provides a public palette for community expression. It is an interactive, highly visible process for integrating curricular initiatives, administrative planning, and sustainability projects. The canvas becomes a template for innovation, imagination, and experimentation, conjuring the art of the possible, linking research and learning to campus infrastructure, while encouraging broad participation.

From Graffiti to Goldsworthies

I'd like to propose a brief portfolio of campus sustainability art projects, with the expectation that they can be implemented as class projects,

campus-wide events, community outreach, alumni reunions, service projects, or any other gatherings that bring people to campus. In many cases,
these can be supervised by either faculty or community artists. They are
ideal venues for short-term residencies.

Sustainability Graffiti Art

Graffiti art is an interesting expression of urban place.[13] Although it is
often denigrated as vandalism, it is a dynamic, talent-rich genre with an
extraordinary diversity of artistic forms and styles. Graffiti artists often
work together developing collage-like murals, painting over or "editing"
other work. There is an expectation that the canvas (which might appear anywhere) will metamorphose. Many cities promote specified public
places as "legal" canvases. Yet the "underground" aspect of the art form is
still prominent—graffiti art shows up in unexpected places. Highly skilled
graffiti art is rich in meaning and magnificently provocative. The "spirit" of graffiti art is to liberate public spaces—from subway cars to the
sides of buildings—as venues for individual expression. Obviously, that
will spur controversy, from the artistic statement to issues of public and
private access. The genre promotes the unexpected. However, providing
some structure to canvas locations promotes graffiti art as more enduring
and thereby enhances its status.

Sustainability graffiti art may take many forms. It may involve more
traditional spray paint compositions with sustainability themes and messages. Or it might entail the use of sustainable materials—found objects,
grasses, moss, or sticks. Or it might involve the combination of both approaches. I recommend that campuses designate a variety of potential
graffiti locations and encourage their use. Depending on which spots are
chosen, who gets involved, and how the space is appropriated, sustainability graffiti art can have curricular applications, stimulate competitions
and contests, promote special events, and generate college fund-raisers.

Sculptures Made with Recycled Materials

Consider using recycled materials as a primary medium for campus
sculptures. There are countless ways to accomplish this, depending on
the waste stream of the campus. You can work with facilities and campus
planners to organize formal class projects. Or you can set up stations and
locations on campus that provide materials and instruction, culminating

in more spontaneous projects. The use of recycled materials is a very interesting way to combine substantive sustainability messages with creative expression. A brief survey of web images with the term "recycled material sculptures" will yield some eye-popping possibilities.[14]

Sustainable Landscaping Art

Both urban and rural campuses have wonderful landscaping opportunities, especially by wisely utilizing green spaces and watercourses. Depending on the specifics of biogeography, topography, and habitat, campus landscaping can serve both instructional and aesthetic purposes. For the most part, campus landscaping is staid and traditional. With some imagination and flexibility you can initiate very interesting sustainable landscaping projects, including the use of gardens, food growing, native plant sculptures, wildflower art, permaculture, composting, and the use of local materials.[15]

Soundscape Designs

You can also construct environments that highlight, orchestrate, and identify sounds. Acoustic ecology emphasizes the importance of linking sound and landscape in order to promote a deeper appreciation of the natural world. The field of soundscape design, pioneered by R. Murray Shafer, integrates an awareness of local sound with architectural planning, musical composition, landscaping, and listening.[16] This is an interesting, largely unexplored, and potentially musical way to engage a campus in further exploring the intricacies of place. It is linked to the sustainability agenda by encouraging sound awareness as crucial to wellness, aesthetics, and interpretation. You can organize places on campus that feature soundscape design, careful listening, or improvisational music making using local materials.[17]

Goldsworthies

In his imaginative science-fiction novel *2312*, Kim Stanley Robinson portrays images of futuristic landscapes on terraformed planets and asteroids throughout the solar system. In this vision, "goldsworthies" are a prominent art form. Named after the great sculptures of Andy Goldsworthy, these are artistic/acoustic compositions that use natural materials to inspire, amplify, and exemplify significant landscape features. They highlight patterns, waves, spirals, and other forms as an interactive, reciprocal process between the artist and the landscape. These range in scale from topographic-feature size creations to microenvironment stone sculptures. They are a brilliant combination of composition and improvisation, yielding an attitude of playing with the landscape, while also calling attention to its magnificence of form.[18]

Using the outdoor landscape as an intentional gallery is neatly adaptable in campus settings, incorporating any of the approaches described above, ranging from whole campus system artwork to nooks, crannies, and corners. It may be commissioned work, or more ephemeral, spontaneous projects. These proposed "goldsworthies" can have a range of interpretive application, from biospheric patterns to the daily habits of sustainable living. As a campus design challenge, it is instructive to consider and utilize "goldsworthies" as intrinsic to all aspects of campus planning.

A Sustainability Aesthetic

What does it mean to link the words "sustainability" and "aesthetic" together? "Aesthetics" refers to a conception of elegance, beauty, and the sublime. Etymologically (from the Greek), it refers to things perceptible by the senses. Philosophically, it is concerned primarily with the extent to which beauty (or ugliness) is inherent or whether it is in the eye of the beholder, and the moral implications of such distinctions. Attaching "sustainability" to "aesthetic" implies that our conception of what is elegant and beautiful is informed by sustainability principles, which are in turn derived from ecological patterns and processes, and ultimately biospheric processes. Sustainability aesthetics are guided by the extent to which human design, architecture, and art emulate the forms and patterns of life processes, the ecosystem and the biosphere.[19]

A sustainability aesthetic is informed by extrinsic and intrinsic influences. By "extrinsic," I refer to how values derived from the sustainability ethos contribute to one's perception of what is appealing. There is an emerging practice of sustainability routines and behaviors that espouses simplicity, conservation, frugality, durability, and resilience as life-enhancing virtues. When determining whether an object, an artifact, a building, or machine is elegant or beautiful, we may assess its design, functionality, and efficiency from a sustainability perspective. How much energy is used? What is it made out of? Where is it made? What kind of resource extraction is entailed? How long may it last? These extrinsic factors do not in themselves necessarily reflect the beauty and elegance of the object, but they are qualities that may inform our aesthetic preference—why we might find a Prius more aesthetically appealing than a Hummer, or a compact solar home more appealing than a sprawling trophy mansion.

By "intrinsic factors," I refer to how aesthetic appeal is informed by patterns in nature. There is an emerging literature and practice of art, design, and architecture whose aesthetic is defined by ecological patterns and processes. It perceives these patterns as inherently beautiful and worthy of emulation, interpretation, and inspiration. This prevailing ecologically informed aesthetic (as demonstrated in expressions as diverse as nature photography, landscape painting, and "green architecture") espouses a convergence between the built environment and the natural world. Further, it suggests that there are reiterative forms and patterns in nature that sustain life, promote human happiness, and appear at multiple perceptual scales.

Christopher Alexander, in his comprehensive four-volume work *The Nature of Order*, proposes fifteen fundamental properties that reflect vitality and life in both buildings and the natural world. He organizes dozens of photographs and illustrations demonstrating their depth and prevalence. The implication is that the emulation of these properties leads to life promoting architecture and building. In brief, the properties are levels of scale, strong centers, boundaries, alternating repetition, positive space, good shape, local symmetries, deep interlock and ambiguity, contrast, gradients, roughness, echoes, the void, simplicity and inner calm, and not-separateness. Alexander is creating an epistemology of natural form that implies both virtue and beauty, connecting patterns in nature to the process of creating life-enhancing built environments. His work uses both intrinsic and extrinsic influences to inform an ecological architecture. Alexander suggests that the "fundamental properties" are embedded in natural processes.[20] Yet his ability to discern these "properties" is deeply informed by his values regarding quality of life and human happiness.

A sustainability aesthetic is informed by the interface between these extrinsic and intrinsic factors. The biomimicry concept, as espoused by Janine Benyus, illustrates this convergence well:

Biomimicry is learning from and then emulating natural forms, processes, and ecosystems to create more sustainable designs. . . . Beauty is a large part of why biomimicry resonates. . . . Having spent 99.9% of our planetary tenure woven deep into the wild, we humans naturally admire the weaverbird's nest, the conch's shell, the scales of a shimmering trout. In fact, there are few things more beautiful to the human soul than good design. When it is good in all aspects—stirring to the senses, fit for its function, elegant in its material choice, and gentle in its manufacture, we can't help but feel delight and the desire to do at least as well in our next design.[21]

Yet not everything in nature is a brilliant design or embodies ecological fitness perfectly. Nature is also replete with invasive species, geological catastrophes, ecological inefficiencies, and evolutionary dead-ends. I, too, am inspired and delighted by convergent, multi-scalar, ecological and geological patterns and processes. However, I am predisposed to find such patterns beautiful because I have learned to appreciate them. I like to think they reflect some deeper biospheric principles. Yet I know, too, that they are informed by a specific cultural perspective, and I am wary of proclaiming anything more than that.[22]

Are Wind Turbines Beautiful, or Ugly?

Many different aspects of culture, background, and personal experience influence style and taste. We can never completely understand or articulate the bases of our aesthetic judgments. Yet it is important to recognize the extent to which our values predispose aesthetics. In the case of sustainability, what we know about the context (resource extraction, toxicity, ecological footprint) matters greatly.

The controversies surrounding wind energy are an interesting example of how a sustainability aesthetic is very much in the eye of the beholder. What is your impression of wind turbines? Are they beautiful or ugly? Sustainability advocates may have very different views on this matter. Wind turbines are typically located in prominent places. They have to be tall to catch the wind. Is it appropriate to put wind turbines on a remote mountain ridge, or in a desert valley, or scattered across the landscape, or 20 miles off the coast of Nantucket? Are they more or less beautiful in one place or another? Do you find them beautiful in smaller numbers, let's say a few wind turbines on a remote hillside powering a farm? How about en masse, let's say thousands of turbines lining an Interstate highway? Are they beautiful when you spot them from an airplane window when you fly over Germany? How do they look on a college campus?

We can apply the same set of questions to solar panels. How do they look in small residential settings as opposed to a giant mirror in orbit around the earth? What about nuclear power plants? When you see a cooling tower, do you immediately think about toxic radiation, or do you see reduced greenhouse-gas emissions? I raise these questions because it is important to recognize the extent to which a sustainability aesthetic is informed by scale, place, and function. Are wind turbines OK until you see them from your house? Or until you hear them when you're trying to sleep at night? Or until they reduce the value of your property?

I happen to find wind turbines elegant and beautiful. Is it because there is something intrinsic in their form, or is it because they represent a way of directly harnessing the power of the biosphere without using fossil fuels? But I find power lines across a landscape terribly ugly. They resemble the giant robots in the 1953 film *War of the Worlds*. When I see power lines I don't necessarily distinguish the source of the energy that they carry. So there is an intrinsic aesthetic challenge in designing more elegant power lines *and* wind turbines. There may well be a design aesthetic that can be applied to both structures that enable them to more appropriately fit the landscape.

If you ever travel by train through Austria or Germany, you might look out the window and see a beautiful waste incinerator or power plant. The works of the remarkable Austrian visual artist Friedensreich Hundertwasser included paintings, postage stamp designs, building facades, architectural designs, and industrial facilities. One of his best-known designs is the domestic rubbish incineration and urban heating unit of the Vienna City Council. Pierre Restany describes it this way:

Hundertwasser succeeded in converting a thankless, unattractive heap of industrial volumes cluttered with pipes and metal-bearing frames, into a mosque-palace. . . . The cathedral-bazaar of sacred shit, surmounted by a minaret-column barred at mid-height by a luminous onion dome, is the most magical signal at the entry to a large city that can be imagined. . . . Faithful to his naturist principles, Hundertwasser made sure that the purification of toxic gasses at the power plant was conducted on a par with his exterior renovation work.[23]

It isn't easy to define what is beautiful and elegant, and it is even more difficult to explain. The more important issue is how and whether the art of sustainability can change the way we see the world.

Can Art Change the Way We See the World?

The sustainability ethos asks us to reexamine how we live and the way we see the world. It challenges us to incorporate biospheric principles and ecological processes as fundamental to such reexamination. If we apply this precept to the idea of a sustainability aesthetic, especially as relevant to college campuses, what role might the arts play? In this chapter I have covered how art enhances biospheric perception, why it is important to cultivate the imagination, what it means to engage in creative sustainability, how to implement various approaches to public art, and how our values contribute to what we think is elegant and beautiful. From a practical perspective, sustainability art improves the quality of campus life, has the potential to enhance our understanding of the basic principles of sustainability, and facilitates collaboration and community. Ultimately, sustainability art promises a deeper awareness of how we understand our relationship to nature.

David Rothenberg, in his magnificent book *Survival of the Beautiful*, considers how the creative process, the making of art, enriches our experience of nature. "I believe the most beautiful art," he writes, "is that which makes the world appear richer, deeper, and more meaningful, making nature seem ever more intricate, interesting, and deserving of our attention and love. There is meaning in nature far beyond use; there is form and beauty far beyond function."[24] Here is Rothenberg's most essential point: "I am most impressed by art that changes the way we see the world."[25]

Rothenberg's book is a wonderful, educationally enriching guide for establishing a sustainability aesthetic on a college campus. The creative process is reciprocal, a collaborative relationship between artists and the community. The art of sustainability is most interesting, effective, and evocative when it allows all participants to gain insight and inspiration in unexpected and revealing ways. If education is ultimately a search for meaning, then art is a creative pathway intrinsic to that search. "Art," Rothenberg writes, "evokes, comments, eludes analysis, takes a stab, and stuns us with its surprising tack. The rules that tell us how to make it are meant to be assimilated and then ignored: they are challenges for the next generation to jump ship from the old. The goal is to change the way we see the world through astonishment and delight."[26] Rothenberg's presentation is embodied in a multi-faceted, textured discussion of art, science, and evolution. His view is that art cannot simply be explained as an adaptive mechanism linked to sexual selection, reduced to a narrow evolutionary view. Rothenberg also explores the bowerbird's remarkable architectural nest designs, the vivid complexity of whales' songs, and the intricacies of animal camouflage. He seeks to integrate a mutually reinforcing spiral of scientific explanation and artistic interpretation, suggesting that, just as the wonders of science inspire art, the work of great artists inspires scientists. Ultimately, the integration of art and science enhances our capacity to perceive beauty. Perceiving, appreciating, creating, and contemplating beauty is essential to human flourishing.

So, yes, art can change the way we see the world because it challenges us to find beauty in unexpected places. The ramifications of that discovery generate insight and inquiry. Supporting artistic expression is an appropriate way for a college campus to address its educational mission, for all subjects and purposes, but certainly in regard to sustainability initiatives, ecological processes, and biospheric principles. This leads to the ultimate rationale for sustainability art—it changes the way we see the world, it makes the world more meaningful, it provokes astonishment and delight, it inspires scientific inquiry, and it encourages human flourishing.

Afterword

Humanity is at a crossroads without historical precedent. As a result of the extraordinary and exponential growth of populations and the expansive dynamic of industrial capitalism, humans are a pervasive and dominant force in the health and well-being of Earth and its inhabitants. Humans are now a planetary force comparable in disruptive power to an Ice Age or an asteroid collision. There is great progress in environmental protection, yet all living systems are in long-term decline and are declining at an increasing rate. We are severely disrupting climate stability. There are huge worldwide social, economic, and public-health challenges, and 25–30 percent of the world's people are consuming 70–80 percent of the world's resources.

How do we ensure that current and future humans will live in strong, secure, thriving, and healthy communities in a world that will have 9 billion people and that plans to increase economic output severalfold by 2050? This is the greatest moral, intellectual, and social challenge that human civilization has ever faced.

"Business as usual" will not work. We need a transformative shift in how we think and act. Higher education has an opportunity to lead the way. Why? Higher education prepares most of the professionals who develop, lead, manage, teach, work in, and influence society's institutions, including the most basic foundation of elementary, middle-school, and secondary education. Society looks to higher education to solve current problems, anticipate future challenges, develop innovative solutions, and model the actions and behaviors that promote those solutions. College and university campuses are social microcosms—miniature cities and regional communities. They are anchor institutions for social and economic development. The 4,500 institutions of higher education in the United States are significant economic engines, with annual operating budgets totaling $325 billion annually. This is approximately 2.5 percent of United States' gross domestic product, and it is greater than the GDP of all but 31 countries. Higher education has the ability to create new and better markets for goods and services that will improve society in multiple ways.

Knowledge is the essence of human adaptive capacity, and higher education is the most prominent knowledge-organizing institution. Higher education can provide a long-term strategic focus, providing the skills and knowledge for business leaders, farmers, scientists, health professionals, psychologists, economists, urban planners, policy analysts, artists, cultural and spiritual leaders, teachers, journalists, advocates, activists, and politicians.

As Mitchell Thomashow writes, higher education's rapidly expanding response to this challenge over the last two decades is a beacon of hope in a sea of turbulence. The encouraging news is the unprecedented growth of distinct academic programs related to sustainability in higher education, especially in the last decade. There are new, innovative sustainability studies and graduate programs in every major scientific, engineering social science, and humanities discipline, as well as in the disciplines of business, law, public health, and religion.

Campus sustainability initiatives are equally impressive. Higher education is embracing programs for water and energy conservation, renewable energy, waste minimization, recycling, "green" building and purchasing, alternative transportation, and local and organic food growing. In the United States, the student sustainability movement is well organized, extensive, and sophisticated; indeed, it may be the most significant student-initiated momentum since the civil rights and antiwar movements of the 1960s. Higher education's sustainability efforts are publicly visible to a degree that was unimaginable a decade ago.

However, these efforts are insufficient. In the industrialized world, we are influenced by a myth of human separation from and domination of nature. The successes of material expansion leads to an assumption that Earth's bounty will forever be available and human technological ingenuity will allow us to ignore ecological limits. In many ways, and often unintentionally, higher education is reinforcing many of the unhealthy, inequitable and unsustainable paths that emerge from that worldview. With a few exceptions, sustainability, as an aspiration for society, is not a central institutional goal.

Frank Rhodes, a former president of Cornell University, suggests that the concept of sustainability offers a new foundation for the liberal arts and sciences. It provides a new focus, sense of urgency, and curricular coherence at a time of drift, fragmentation, and insularity in higher education. Sustainability provides hope and opportunity for facilitating institutional renewal and revitalizing higher education's sense of mission.

Mitchell Thomashow shows us what higher education would look like, how it would organize, and how it would act if creating a healthy, just and sustainable human society were its central purpose. He describes the essence of campus transformation in elegant, understandable and practical ways. He specifies the kinds of partnerships with local and regional communities, the private sector, and social organizations that will be necessary. He provides a framework for building on and rapidly expanding higher education's sustainability efforts. The nine elements are the connective thread for campuses to model a thriving civil and sustainable society.

Many critics assume that the ideal of higher education's leading the way toward climate neutrality is unachievable. We must make that which seems impossible *inevitable*. If we continue business as usual, today's students and their children will experience the worst effects of climate disruption, and will have greatly diminished prospects for prosperity, peace, and security. We are facing the greatest intergenerational equity challenge in modern history. Higher education is the last hope for creating a transformational process that addresses these challenges. Its influence is widespread and significant. It has the stature, the expertise, and the intellectual capital to lead the way.

I founded the American College & University Presidents' Climate Commitment in 2006 with the idea that by mobilizing the will and capacity of university presidents we can leverage higher education to demonstrate exemplary sustainability leadership. Richard Cook, president emeritus of Allegheny College and one of the founders of the ACUPCC, posed this challenge well in a private conversation with me:

I liken this pledge [the ACUPCC] to President Kennedy's promise to get men to the moon and back within a decade. Neither he nor a cadre of engineers and scientists knew exactly how this could be accomplished or if, indeed, it could be. But making a bold pledge to accomplish a strategically important end spurred attention, resources, talent, and urgency to a lofty goal that would be difficult to attain. In much the same way, the commitment to becoming climate-neutral institutions will spur development and accountability, and will surely, in most cases, produce more and better results in a shorter period of time than something short of a specific target. The collective voice of higher education can spotlight our sincere concern and commitment to action in ways that few if any other sectors can. We are providing the research that highlights the climate concern; we can also provide many of the solutions. If the colleges and universities don't lead, who will?

Anthony Cortese

Notes

Introduction

1. For a comprehensive overview of this ecological crisis, there are no better assessments than those issued by the International Geosphere Biosphere Program. See especially *Global Change and the Earth System* (available for free downloading at http://www.igbp.net).

2. The respective URLs are http://www.igbp.net, http://www.ipcc.ch, and http://www.globalchange.gov.

3. For a comprehensive overview of sustainability initiatives in a variety of campus settings, see Bartlett and Chase, *Sustainability in Higher Education*. Also see Martin and Samels, *The Sustainable University.*.

Chapter 1

1. The film is available at http://www.roadnottaken.info.

2. See Bill McKibben's "My Road Trip with a Solar Rock Star" and Jean Altomare's "On Anger and Activism" in the 2011 issue of *Hawk & Handsaw*.

3. Pielou, *The Energy of Nature*, p. 3.

4. The seminal works on energy security and political history are Daniel Yergin's *The Prize* and *The Quest*.

5. Andrew Kirk's *Counterculture Green* is an excellent history of the "soft energy" path as linked to the publication of the *Whole Earth Catalog*.

6. Schneider's book, one of the first popular treatments of global warming, catapulted the idea from an obscure Keeling curve to an international concern.

7. For two excellent first-hand accounts by scientists of how their research was challenged, see Hansen, *Stories of My Grandchildren* and Bradley, *Global Warming and Political Intimidation*.

8. For more on bioregionalism, see Lynch, Glotfelty, and Armbruster, *The Bioregional Imagination.*.

9. "The ACUPCC Greenhouse Gas Inventory Brief," at http://www2.presidentsclimatecommitment.org.

10. See, for example, the UCal Berkeley Campus Energy Map (available at http://berkeley.openbms.org).

11. The annual (as of 2012) Behavior, Energy, and Climate Conference is of interest. More information about the conference can be found at http://beccconference .org.

12. For more on Unity House, see Koones, *Prefabulous*.

13. Thomas Friedman discusses the energy Internet on pp. 224–236 of *Hot, Flat, and Crowded*.

14. Lovins, *Reinventing Fire*, p. 167.

15. See, for example, the Wikipedia article on depreciation (http://en.wikipedia .org/wiki/Depreciation).

16. Lovins, *Reinventing Fire*, pp. 87–88.

17. Ibid., p. 89.

18. Ibid., pp. 117–118.

19. Ibid., p. 169.

20. A good way to review these "transformational approaches" is to read some of the better climate action plans submitted to the ACUPCC. They can be found at http://rs.acupcc.org/stats/caps-neutrality/. The CAPs for Arizona State University, Carleton College, and New York University are good starting points.

Chapter 2

1. More on the Maine Organic Farm and Growers Association can be found at http://www.mofga.org.

2. For more on the link between rigorous observations and ecological methodology, see Sagarin and Pauchard, *Observational Ecology*.

3. Darwin, *The Voyage of the Beagle*, p. 284.

4. Pollan, *The Omnivore's Dilemma*, p. 7.

5. Nabhan, *Desert Terroire*, p. 11.

6. In the late 1960s Euell Gibbons wrote numerous popular guides to finding wild edible foods.

7. Pollan, *The Omnivore's Dilemma*, pp. 10–11.

8. For more on the relationship between diet and health, see the online journal *Nutrition and Diabetes* (http://www.nature.com/nutd/index.html).

9. In *Superimmunity*, Joel Fuhrman presents a persuasive argument in support of raw foods and the advantages of a high-nutrition diet.

10. On sugar, salt, and fat vigilance, see Moss, *Salt, Sugar, Fat*. For some great common-sense food advice, see Pollan, *In Defense of Food.*.

11. Your observations will also reveal a great deal about campus multi-culturalism (see Tatum, *Why Are All the Black Kids Sitting Together in the Caféteria?*). Food choices are relevant to cultural and racial identity.

12. For more on Middlebury College's food program, see http://www.middlebury .edu/sustainability/food/dining. Peggy Bartlett's report on the Emory University program (available at http://www.aaanet.org) is framed from an anthropological perspective.

13. For more on Real Food Challenge, see http://www.realfoodchallenge.org.

14. For more on the University of Georgia's food service, see http://foodservice .uga.edu/about/sustainability.

15. Some colleges will insist that a food service meet sustainability criteria when they negotiate a contract with it. See, for example, http://articles.mcall.com/2013 -05-14/news/mc-lafayette-food-service-20130514_1_sodexo-healthy-food -compass-group.

16. On how gardens inspire an ethic of care, see Nollman, *Why We Garden.*

17. On the various functions of Seattle University's urban gardens, see http:// www.seattleu.edu/facilities/inner.aspx?id=35748&utm_source=homepage&utm _medium=CampusGardens&utm_campaign=Centerpieces. On how Messiah College's gardens contribute to biodiversity, see http://www.messiah.edu/sustainability/ stewardship/biodiversity.html. On how Stockton College incorporates campus biodiversity and gardens into all aspects of its master planning process, see http:// intraweb.stockton.edu/eyos/page.cfm?siteID=172&pageID=12.

18. See page 53 of Menzel and D'Aluisio, *Hungry Planet.*

19. James Farrell's book *The Nature of College* is instructive, especially the chapter titled "Food for Thought."

20. On the UCSC Food Systems Working Group, see http://casfs.ucsc.edu/farm-to -college/how-to-get-involved.

Chapter 3

1. William Leiss offers a brilliant critique of affluence in *The Limits to Satisfaction.*

2. Barry, *Getting Better*, pp. 65, 67, and 70.

3. Shirky, *Cognitive Surplus.*

4. Nielsen, *Reinventing Discovery.*

5. Diamandus and Kotler, *Abundance.*

6. The Global Footprint network offers good curricular resources for this type of inquiry at http://www.footprintnetwork.org.

7. The STARS technical manual can be found at http://stars.aashe.org.

8. The quotation from Paul Anastas is from an e360 interview (available at http:// e360.yale.edu).

9. For more on this history, see Souder, *On a Farther Shore..*

10. Planet Green's "green materials guide" can be found at http://planetgreen .discovery.com.

11. This quotation is from the Green Chemistry website of the U.S. Environmental protection Agency (http://www.epa.gov/greenchemistry/). Also see Anastas and Warner, *Green Chemistry*.

12. For the principles of "green chemistry," see http://greenchemistry.yale.edu/ javascript/tinymce/plugins/filemanager/files/principles-of-green-chemistry.pdf.

13. The source of the quotation on "green chemistry" is http://www.epa.gov/ greenchemistry/.

14. For more on the concept of "restore," see Shedroff, *Design Is the Problem.*.

15. Church and Regis, *Regenesis*, p. 2.

16. Ibid., p. 7.

17. Ibid., p. 9.

18. For an outstanding arts-oriented guide to some of the exciting innovations in biodesign incorporating the ideas of "restore," "regenerate," and "redesign," see Myers, *Biodesign*.

19. More information on the projects at Pratt's Center for Sustainable Designable Strategies is available at http://csds.pratt.edu.

20. For a reasonably complete list of sustainability-related academic programs, see http://sspp.proquest.com/sspp_institutions/display/universityprograms.

21. See The Designers Field Guide to Sustainability at http://www.lunar.com/ docs/the_designers_field_guide_to_sustainability_v1.pdf. Also see *Greensource: The Magazine of Sustainable Design* at http://greensource.construction.com.

22. On durability and design, see http://www.durabilityanddesign.com.

23. On resilience, see http://www.resilientdesign.org/.

24. On modularity and sustainable design, see http://www.empf.org/empfasis/ sept03/designforsus.htm.

25. On adaptability, see http://www.structuremag.org/article.aspx?articleID =1040.

26. On "green ergonomics," see http://www.indesignlive.com/articles/in-review/ report/Exploring-Green-Ergonomics#axzz2BYNur88t.

27. On craft and sustainability, see http://www.craftscouncil.org.uk/about -us/press-room/view/2011/craft-environmental-sustainability-201101261052 -4d3ffd03de9c6?from=/about-us/press-room/.

Chapter 4

1. These qualities are derived from the United Nations' excellent ESCAP (Economic and Social Commission for Asia and the Pacific) list, available at http:// www.unescap.org. Also see Peter Brown's book *Right Relationship*, on page 110 of which the following criteria for good governance are listed: capacity and authority, credibility, accountability and effectiveness, transparency, subsidiarity.

2. For a discussion of the implications of the Anthropocene concept, see http:// dotearth.blogs.nytimes.com/2012/09/17/the-anthropocene-as-environmental -meme-andor-geological-epoch/. Also see the neatly illustrated webpage http://

www.anthropocene.info/en/home and the National Geographic popularization (http://ngm.nationalgeographic.com/2011/03/age-of-man/kolbert-text).

3. There is an emerging literature describing this process. See especially Bartlett and Chase, *Sustainability in Higher Education.*.

4. In *The Necessary Revolution*, Peter Senge integrates change management and sustainability. For an excellent guide to higher-education leadership, especially regarding change management and "turnaround challenges," see Fullan and Scott, *Turnaround Leadership for Higher Education.*.

5. See Goleman and Boyatzis, "Social Intelligence and the Biology of Leadership."

6. Csikszentmihalyi, *Good Business.*

7. We are just now beginning to see interesting case studies of how colleges and universities embody such transformational change. See Bartlett and Chase, *Sustainability in Campus* and Martin et al., *The Sustainable University*. The variety of institutional situations and histories, and issues of scale and culture limit our ability to generalize. However, in almost all cases, whether reviewing the case of a large state system (Portland State University), a large private university (Cornell), community colleges (Butte College), or smaller state and privates (the University of Maine at Farmington, Allegheny College), to name just a few, you will find that coherent initiatives come from an aligned senior team.

8. Professor Mick Womersley spearheaded many of these efforts. His creative, irreverent, and poignant blog Sustainability Thought and Deed can be found at http://ucsustainability.blogspot.com.

9. Unity House was a prototype for a new style of "green" residential dwellings. See http://unityhomes.com.

10. If you are interested in a deep understanding of the social and emotional dimensions of change management, it is very helpful to look within and nourish the capacity to understand your own emotional waves and cycles. I found two "classic" texts particularly insightful. Pema Chodron's book *No Time to Lose* provides an interesting Buddhist perspective on balancing emotional demands, and Jack Balkin's translation of the *I Ching* is very helpful for scaling leadership issues with the vicissitudes of human behavior.

Chapter 5

1. The definition is from http://www.britannica.com.

2. Boulding, *Ecodynamics*, p. 174.

3. The definition is from http://www.naturalcapitalproject.org/.

4. Source: http://www.eoearth.org/article/Natural_capital.

5. Source: http://www.bettertogether.org/socialcapital.htm.

6. Source: http://www.cpn.org/tools/dictionary/capital.html.

7. Source: http://www.businessdictionary.com/definition/intellectual-capital.html.

8. On measuring intellectual capital, see http://www.skyrme.com/insights/24kmeas.htm.

9. For an excellent introduction to strategic budgeting in higher education, and a book that covers these ratios, see Chabotar, *Strategic Finance*. The ratios are also described at http://regents.ohio.gov/financial/campus_accountability/index.php.

10. See http://www.socialcapitalresearch.com/measurement.html.

11. For more on ecosystem services, see the Millennium Ecosystem Assessment, *Ecosystems and Human Well-Being*. The following passage on page 155) is relevant here: "The simplest form of ecosystem valuation for economists is to hold that an ecosystem has a value equivalent to its ecological yield valued as it would be on commodity markets: for the value of water, wood, fish or game, that is purified or nurseried or generated or harboured in that ecosystem. Thus, a price can be put on the natural capital of an ecosystem based on the price of natural resources it yields each year."

12. In 2012, my successor as president convinced Unity College to divest itself of all the oil-company stocks in its endowment. This was a bold statement of values. There is now an active and controversial movement, spearheaded by students, to convince their schools to do likewise. Obviously this divestment challenge brings all kinds of political issues to the foreground of campus affairs. I view it as an outstanding educational opportunity to discuss values and investment, or the true value of investment.

13. For more on the Sustainable Endowments Institute, see http://www.greenreportcard.org/report-card-2011/executive-summar.

14. For more on the report card, see http://www.greenreportcard.org/report-card-2011/indicators.

15. Helpful resources are available at http://www.ghgprotocol.org.

16. For more on campus climate action planning, see http://www.presidentsclimatecommitment.org/resources/climate-action-planning.

17. The Carleton College plan is available at http://apps.carleton.edu/sustainability/about/cap/.

18. The Arizona State CAP is available at http://carbonzero.asu.edu/CarbonPlan022410.pdf.

19. Esty and Simmons, *The Green to Gold Business Playbook*, p. 305.

20. See http://en.wikipedia.org/wiki/Cradle_to_Cradle:_Remaking_the_Way_We_Make_Things.

21. The quotation on life-cycle assessment is from http://en.wikipedia.org/wiki/Life_cycle_assessment.

22. See pp. 101–137 of Anderson, *Mid-Course Correction*.

23. Esty and Simmons, *The Green to Gold Business Playbook*, pp. 81–109.

24. On STARS, see http://www.aashe.org/files/documents/STARS/stars_1.1_administrative_update_one_technical_manual.pdf.

25. On the University of Exeter's One Planet MBA program, see http://business-school.exeter.ac.uk/mba/.

26. Harvard University's excellent life-cycle costing calculator is available at http://www.green.harvard.edu.

27. For an impressive group of revolving loan fund case studies, see http://greenbillion.org/resources/.

28. For a full list of these colleges and brief profiles, see http://greenbillion.org/participants/.

29. All the documents referred to in the preceding paragraph, and this one, can be found at www.greenbillion.org.

30. See, for example, http://view.fdu.edu/default.aspx?id=5209.

31. More information on TCCPI is available at http://www.tccpi.org/.

32. Source: http://sustainability.asu.edu/index.php.

Chapter 6

1. Owen Flanagan, in his book *The Really Hard Problem,* provides a philosophical methodology that seeks to better understand the concept of human flourishing.

2. Two excellent texts that elaborate on the relationship between human health and ecosystem processes are *Ecosystem Health*, edited by Rapport et al., and *Sustaining Life*, edited by Chivian and Bernstein.

3. See the *Chronicle of Higher Education*'s annual "Great Colleges to Work For" studies, available at http://chronicle.com.

4. The definition of community vitality is from http://oregonexplorer.info/rural/.

5. See Torres, Jones, and Renn, "Identity Development Theories in Student Affairs: Origins, Current Status, and New Approaches.

6. For a good overview of the concept of ecological resilience, see http://en.wikipedia.org/wiki/Resilience_(ecology).

7. Davidson, *The Emotional Life of Your Brain..*

8. For an excellent definition of resilience, reflecting its relevance for both human and ecological systems, see Jean-Claude Laprie, "From Dependability to Resilience" (available at http://www.ece.cmu.edu).

9. See Galluzi, Eyzaguirre, and Negri, "Home Gardens: Neglected Hotspots of Agro-biodiversity and Cultural Diversity."

10. See J. Pretty et al., "The Intersections of Biological Diversity and Cultural Diversity: Towards Integration," available at http://www.conservationandsociety.org.

11. The UNESCO document is available at http://unesdoc.unesco.org. The passage cited is on page 7.

12. This challenge may be increasingly linked to climate adaptation. What role will college and university campuses play in housing dislocated students, climate refugees from other places? This was an important role for many campuses during Hurricane Katrina, in 2005.

13. Source: Kellert, *Building for Life*, p. 5.

14. Ibid.

15. Ibid., p. 139.

16. Ibid., p. 176.

17. The President's Higher Education Community Service Honor Roll, launched in 2006, annually highlights the role colleges and universities play in solving community problems and placing more students on a lifelong path of civic engagement by recognizing institutions that achieve meaningful, measurable outcomes in the communities they serve. The URL is http://www.nationalservice.gov/about/initiatives/honorroll.asp.

18. The NSSE website has the latest results of its annual surveys. They are available at http://nsse.iub.edu

19. The Maine Campus Compact study "The Effect of Service-Learning on Retention" is available at http://www.compact.org.

20. Source: Ralph Waldo Emerson, "The American Scholar" (available at http://www.emersoncentral.com).

21. Source: Cafaro, *Thoreau's Living Ethics*, p. 30.

22. Source: Shi, *The Simple Life*, p. 176.

23. Cafaro eloquently poses this challenge on p. 86 of *Thoreau's Living Ethics*: "The relative value of pleasure, self-development, and personal achievement in a good human life is perhaps the most vexing question in virtue ethics. We may strive to further all these, while recognizing that sometimes we must choose between them. Each of us answers this question within his or her own life."

24. Jim Dodge tells this story in his foreword to *The Gary Snyder Reader*.

25. Cafaro, *Thoreau's Living Ethics*, p. 156.

Chapter 7

1. For more details, see https://www.unity.edu/academics/centers.

2. Bowers' book *University Reform in an Era of Global Warming* is a controversial discussion of how curriculum reproduces culture.

3. On using sustainability to transform an entire curriculum at a Research One institution, see Aber, Kelly, and Mallory, *The Sustainable Learning Community*.

For an excellent anthology that covers both curricular and infrastructure transformation, see Martin et al., *The Sustainable University*.

4. This idealized discussion, although representative of many, represents a discussion among college presidents at the ACUPCC summit at American University in June of 2012.

5. Barber, *Strong Democracy*, pp. 180–182.

6. For an excellent and original discussion of all the ways everyday life on a college campus can be linked to sustainability, see Farrell, *The Nature of College*.

7. As of this writing, Arizona State University, Portland State University, Unity College, Green Mountain College, Lane Community College, and the Col-

lege of Lake County are representative of how presidential leadership can affect curricular change. For a more comprehensive list, see http://www .presidentsclimatecommitment.org/ or https://stars.aashe.org/.

8. As of this writing, some of the most prominent are the AASHE conference, Powershift, Smart and Sustainable, the California Higher Education Sustainability Conference, and scores of regional gatherings.

9. See the National Wildlife Federation's database of innovative campus sustainability projects at http://www.nwf.org/Campus-Ecology/Campus-Search.aspx.

10. To get a good sense of the extent of these programs, visit http://www.aashe .org/connect/enewsletters/bulletin.

11. See, for example, the advice on promoting sustainable behavior at http:// sustainability.berkeley.edu/os/pages/talkinglouder/docs/Promoting_Sustain_Be- havior_Primer.pdf. Much of that advice is derived from Doug McKenzie-Mohr's book *Fostering Sustainable Behavior* and the related website http://www.cbsm .com/pages/guide/preface/.

12. For more on these nine principles, see http://www.exploratorium.edu/IFI/ resources/research/constructivistlearning.html.

13. Berea College (in Kentucky) and Warren Wilson College (in North Carolina) are two work colleges with excellent sustainability programs. For more on the idea of the work college, see http://www.workcolleges.org/member-colleges.

14. Source: Bowers, "The Challenge of Making the Transition from Individual to Ecological Intelligence in an Era of Global Warming," p. 26.

15. See Daniel Goleman, "What Is Ecological Intelligence?" at http://danielgoleman .info.

16. A fine starting point for reviewing these efforts is http://greenschools.net/ article.php?id=246.

17. David Orr's book *Ecological Literacy* set the pace for these efforts.

18. Source: Thomashow, "The Gaian Generation."

19. Source: Kahneman, *Thinking Fast and Slow*, pp. 4 and 45.

20. Davidson, *The Emotional Life of Your Brain*, p. xvii.

21. Several works stand out: Hall, *Wisdom*; Schultz, *Being Wrong*; Johnson, *Where Good Ideas Come From*; Brockman, *This Will Make You Smarter.*

Chapter 8

1. The Stanford Encyclopedia of Philosophy (http://plato.stanford.edu) has a helpful article on phenomenology.

2. The work of the philosopher David Abram is exemplary. In *The Spell of the Sensuous* he explores the phenomenological origins of environmental perception, and in *Becoming Animal* he brilliantly applies these concepts to direct experience. For a "handbook-style" approach, see Beck and Cable, *Interpretation for the 21st Century* and Serrell, *Exhibit Labels.*

3. See http://www.storyofstuff.org/.

4. Tyler Volk's book *CO₂ Rising* traces the movement of a carbon atom through geological space and time. It is highly recommended as an interpretive template for utilizing this shifting-scale perspective.

5. See chapter 3 of Thomashow, *Bringing the Biosphere Home*.

6. Snyder, *The Practice of the Wild*; Orr, *Ecological Literacy*; Sobel, *Childhood and Nature*; Chawla, *In the First Country of Places*; Basso, *Senses of Place*.

7. For an excellent compilation of essays attesting to the educational necessity of natural history, see Fleischner, *The Way of Natural History*.

8. On the relationship among identity, place, and meaning, see Thomashow, *Ecological Identity*; Clayton and Opotow, *Identity and the Natural Environment*; Deming and Savoy, *The Colors of Nature*.

9. Thomashow, *Bringing the Biosphere Home*.

10. For more ideas on making the geological time scale tangible, see http://longnow.org/clock/.

11. See http://www.inaturalist.org/.

12. See http://www.google.com/earth/index.html.

13. For more about bioblitzes, see http://education.eol.org/bioblitz.

14. See Pietsch, *Trees of Life*.

15. See Louv, *The Nature Principle*.

16. Lewis Mumford's classic work *The Culture of Cities* is a brilliant discussion of the relationship between the rural and the urban as manifested in architecture and planning. Christopher Alexander's *A Pattern Language* is a classic of ecologically oriented landscape and building design. Alexander's more recent four-volume opus *The Nature of Order* is a virtual encyclopedia of such approaches, embedded in a philosophy of living and learning. Also see Hester, *Design for Ecological Democracy* and Kellert, *Building for Life*.

17. A definition of "living building" can be found at http://thegoodhuman.com. For more on the Living Building Challenge, visit https://ilbi.org.

18. On the educational experience of designing the Lewis Center, see Orr, *Design on the Edge*.

19. For more on the Lewis Center, see http://www.mcdonoughpartners.com/projects/view/adam_joseph_lewis_center_environmental_studies_oberlin_college.

20. For more on Unity House and Terahaus, see http://terrahaus.wordpress.com/2011/07/05/unity-house-and-terrahaus/.

21. See Brand, *How Buildings Learn*.

22. Source: http://www.mcdonough.com.

23. Source: http://www2.coloradocollege.edu.

24. Source: http://www.pdx.edu.

25. Source: http://blogs.ucdavis.edu.

Chapter 9

1. Some of Greg Mort's work can be seen at http://www.gregmorteditions.com.

2. A few examples from the illustrated biosphere library: Bill Atkinson, *Within the Stone,* Yann Artus-Bertrand, *Earth From Above,* Ron Redfern, *Origins.*

3. An excellent discussion of the concept of imagination can be found at http://csmt.uchicago.edu/glossary2004/imagination.htm. For a comprehensive scholarly discussion of the Western intellectual history of imagination, see Engell, *The Creative Imagination.*

4. Source: Johnson, *Where Good Ideas Come From,* pp. 16–17.

5. Ibid., p. 19.

6. Ibid., p. 20.

7. The "seven patterns" are the adjacent possible, liquid networks, the slow hunch, serendipity, error, exaptation, and platforms.

8. Source: Holmes, *The Age of Wonder,* p. 276.

9. Ibid., p. 469.

10. Nachmanovich, *Free Play,* p. 27.

11. What Is Creative Sustainability? is a column in *Hawk & Handsaw.* The passages cited here are excerpted from five issues.

12. I highly recommend the comprehensive review, taxonomy, and explication of eco art as organized and interpreted by Linda Weintraub in *To Life!*

13. See, for example, Ganz, *Graffiti World.*

14. Patrick Dougherty's environmental sculpture can be seen at http://www.stickwork.net.

15. See http://www.caes.uga.edu/campus/griffin/garden/WildFlowers.htm.

16. See Shafer, *The Soundscape.*

17. For some interesting examples, see http://wfae.proscenia.net.

18. All of Andy Goldsworthy's books are filled with great possibilities. Start with *Time.*

19. For a design perspective shaped by a sustainability aesthetic, see Hosey, *The Shape of Green.*

20. The fifteen fundamental properties are explained and illustrated on pp. 144–242 of the first volume, *The Phenomenon of Life.*

21. See http://biomimicry.net/about/biomimicry/a-biomimicry-primer/.

22. One of the best books on patterns in nature is Ball, *The Self-Made Tapestry.*

23. Restany, *Hundertwasser,* pp. 47–49.

24. Rothenberg, *Survival of the Beautiful,* p. 34.

25. Ibid., p. 57.

26. Ibid., p. 58.

Bibliography

Aber, John, Tom Kelly, and Bruce Mallory. *The Sustainable Learning Community*. University of New Hampshire Press, 2009.

Abram, David. *The Spell of the Sensuous: Perception and Language in a More-Than-Human World*. Random House, 1996.

Abram, David. *Becoming Animal: An Earthly Cosmology*. Pantheon, 2010.

Alexander, Christopher. *The Nature of Order: An Essay on the Art of Building and The Nature of the Universe*. Center for Environmental Structure, 2002.

Alexander, Christopher, Sara Ishikawa, and Murray Silverstein. *A Pattern Language*. Oxford University Press, 2007.

Altomare, Jean. "On Anger and Activism." *Hawk & Handsaw* 4 (2011): 10–12.

Anastas, Paul T., and John C. Warner. *Green Chemistry: Theory and Practice*. Oxford University Press, 2000.

Andersen, Ray. *Mid-Course Correction*. Chelsea Green, 1998.

Artus-Bertrand, Yann. *Earth from Above*. Abrams, 2005.

Atkinson, Bill. *Within the Stone*. BrownTrout, 2004.

Balkin, Jack R. *The Laws of Change: I Ching and the Philosophy of Life*. Schocken, 2002.

Ball, Philip. *The Self-Made Tapestry: Pattern Formation in Nature*. Oxford University Press, 1999.

Barber, Benjamin. *Strong Democracy*. University of California Press, 1984.

Barry, Charles. *Getting Better: Why Global Development Is Succeeding—And How We Can Improve the World Even More*. Basic Books, 2011.

Bartlett, Peggy F., and Geoffrey W. Chase. *Sustainability on Campus: Stories and Strategies for Change*. MIT Press, 2004.

Bartlett, Peggy F., and Geoffrey W. Chase. *Sustainability in Higher Education: Stories and Strategies for Transformation*. MIT Press, 2013.

Beck, Larry, and Ted Cable. *Interpretation for the Twenty-First Century: Fifteen Guiding Principles for Interpreting Nature and Culture*. Sagamore, 2002.

Boulding, Kenneth. *Ecodynamics: A New Theory of Societal Evolution*. Sage, 1978.

Bowers, C. A. "The Challenge of Making the Transition from Individual to Ecological Intelligence in an Era of Global Warming." *Proceedings of the Media Ecology Association* 11 (2010): 21–28.

Bowers, C. A. *University Reforms in an Era of Global Warming.* Eco-Justice, 2011.

Bradley, Raymond S. *Global Warming and Political Intimidation: How Politicians Cracked Down on Scientists as the Earth Heated Up.* University of Massachusetts Press, 2011.

Brand, Stewart. *How Buildings Learn: What Happens After They're Built.* Penguin, 1995.

Brand, Stewart. *Whole Earth Manifesto: An Ecopragmatist Manifesto.* Viking, 2009.

Brockman, John. *This Will Make You Smarter: New Scientific Concepts to Improve Your Thinking.* HarperCollins, 2012.

Brown, Peter G., and Geoffrey Carver. *Right Relationship: Building a Whole Earth Economy.* Berrett-Koehler, 2009.

Cafaro, Philip. *Thoreau's Living Ethics: Walden and the Pursuit of Virtue.* University of Georgia Press, 2004.

Chabotar, Kent. *Strategic Finance: Planning and Budgeting for Boards, Chief Executives, and Finance Officers.* Association of Governing Boards of Universities and Colleges, 2006.

Chalwa, Louise. *In the First Country of Places: Nature, Poetry and Childhood Memory.* SUNY Press, 1994.

Chivian, Eric, and Aaron Bernstein, eds. *Sustaining Life: How Human Health Depends on Biodiversity.* Oxford University Press, 2008.

Chodron, Pema. *No Time to Lose: A Timely Guide to the Way of the Bodhisattva.* Shambhala, 2005.

Church, George, and Ed Regis. *Regenesis: How Synthetic Biology Will Reinvent Nature and Ourselves.* Basic Books, 2012.

Clayton, Susan, and Susan Opotow, eds. *Identity and the Natural Environment: The Psychological Significance of Nature.* MIT Press, 2003.

Crist, Eileen, and H. Bruce Ronker, eds. *Gaia in Turmoil: Climate Change, Biodepletion, and Earth Ethics in an Age of Crisis.* MIT Press, 2010.

Csikszentmihalyi, Mihaly. *Good Business: Leadership, Flow and the Making of Meaning.* Penguin, 2004.

Darwin, Charles. *The Voyage of the Beagle.* Penguin, 1989.

Davidson, Richard J., with Sharon Begley. *The Emotional Life of Your Brain.* Hudson Street, 2012.

Deming, Alison Hawthorne, and Lauret E. Savoy, eds. *The Colors of Nature: Culture, Identity and the Natural World.* Milkweed, 2011.

Diamandus, Peter H., and Steven Kotler. *Abundance: The Future is Better Than You Think.* Free Press, 2012.

Engell, James. *The Creative Imagination: Enlightenment to Romanticism*. iUniverse, 1999.

Esty, Daniel C., and P. J. Simmons. *The Green to Gold Business Playbook: How to Implement Sustainability Practices for Bottom-Line Results in Every Business Function*. Wiley, 2011.

Farrell, James J. *The Nature of College: How a New Understanding of Campus Life Can Change the World*. Milkweed, 2010.

Feld, Steven, and Keith Basso, eds. *Senses of Place*. School of American Research Press, 1996.

Flanagan, Owen. *The Really Hard Problem: Meaning in a Material World*. MIT Press, 2007.

Fleischner, Thomas Lowe, ed. *The Way of Natural History*. Trinity University Press, 2011.

Friedman, Thomas. *Hot, Flat and Crowded: Why We Need a Green Revolution—And How It Can Renew America*. Farrar, Straus and Giroux, 2008.

Fuhrman, Robert. *Superimmunity*. HarperCollins, 2011.

Fullan, Michael, and Geoff Scott. *Turnaround Leadership for Higher Education*. Wiley, 2009.

Galluzi, Gea, Pablo Eyzaguirre, and Valeria Negri. "Home Gardens: Neglected Hotspots of Agro-biodiversity and Cultural Diversity." *Biodiversity and Conservation* 19 (2010), no. 13: 3635–3654.

Ganz, Nicholas. *Graffiti World: Street Art From Five Continents*. Abrams, 2009.

Gardner, Howard. *Frames of Mind: The Theory of Multiple Intelligences*. Basic Bools, 2011.

Goldsworthy, Andy. *Time*. Abrams, 2000.

Goleman, Daniel, and Richard Boyatzis. "Social Intelligence and the Biology of Leadership." *Harvard Business Review*, September 2008: 74–81.

Hall, Stephen S. *Wisdom: From Philosophy to Neuroscience*. Knopf, 2010.

Hansen, James. *Stories of My Grandchildren: The Truth About the Coming Climate Catastrophe and Our Last Chance to Save Humanity*. Bloomsbury, 2009.

Hester, Randolph T. *Design for Ecological Democracy*. MIT Press, 2006.

Holmes, Richard. *The Age of Wonder: How the Romantic Generation Discovered the Beauty and Terror of Science*. Pantheon, 2008.

Hosey, Lance. *The Shape of Green: Aesthetics, Ecology, and Design*. Island, 2012.

Johnson, Steven. *Where Good Ideas Come From: The Natural History of Innovation*. Riverhead, 2010.

Kahneman, Daniel. *Thinking, Fast and Slow*. Farrar, Straus and Giroux, 2011.

Kareiva, Peter, Heather Tallis, Taylor H. Ricketts, Gretchen C. Daily, and Stephen Polasky. *Natural Capital: Theory and Practice of Mapping Ecosystem Services*. Oxford University Press, 2011.

Kellert, Stephen R. *Building for Life: Designing and Understanding the Human-Nature Connection*. Island, 2005.

Kirk, Andrew G. *Counterculture Green: The Whole Earth Catalog and American Environmentalism.* University Press of Kansas, 2007.

Koones, Sherri. *Prefabulous +Almost Off the Grid.* Abrams, 2012.

Leiss, William. *The Limits to Satisfaction: An Essay on the Problem of Needs and Commodities.* University of Toronto Press, 1976.

Louv, Richard. *The Nature Principle: Reconnecting with Life in a Virtual Age.* Algonquin, 2011.

Lovins, Amory, and Rocky Mountain Institute. *Reinventing Fire: Bold Business Solutions for the New Energy Era.* Chelsea Green, 2011.

Lynch, Tom, Cheryll Glotfelty, and Karla Armbruster, eds. *The Bioregional Imagination: Literature, Ecology, and Place.* University of Georgia Press, 2012.

Martin, James, James E. Samels, and Associates. *The Sustainable University: Green Goals and New Challenges for Higher Education Leaders.* Johns Hopkins University Press, 2012.

McKenzie-Mohr, Douglas. *Fostering Sustainable Behavior.* New Society, 1999.

McKibben, Bill. "My Road Trip with a Solar Rock Star." *Hawk & Handsaw* 4 (2011): 7–9.

Menzel, Peter. *Material World: A Global Family Portrait.* Sierra Club Books, 1994.

Menzel, Peter, and Faith D'Aluisio. *Hungry Planet: What the World Eats.* Material World, 2007.

Merkel, Jim. *Radical Simplicity: Small Footprints on a Finite Earth. Gabriola Island.* New Society, 2003.

Millennium Ecosystem Assessment. *Ecosystems and Human Well-Being: Synthesis.* Island, 2005.

Moss, Michael. *Salt, Sugar, Fat: How the Food Giants Hooked Us.* Random House, 2013.

Mumford, Lewis. *The Culture of Cities.* Harcourt, Brace, 1938.

Myers, William. *Biodesign: Nature, Science, Creativity.* Museum of Modern Art, 2012.

Nabhan, Gary. *Desert Terroire: Exploring the Unique Flavors and Sundry Places of the Borderlands.* University of Texas Press, 2012.

Nachmanovich, Steven. *Free Play: Improvisation in Life and Art.* Tarcher/Putnam, 1990.

Nielsen, Michael. *Reinventing Discovery: The New Era of Networked Science.* Princeton University Press, 2012.

Nollman, Jim. *Why We Garden: Cultivating a Sense of Place.* Sentient Publications, 2005.

Orr, David. *Ecological Literacy: Education and the Transition to a Postmodern World.* SUNY Press, 1991.

Orr, David. *Design on the Edge: The Making of a High Performance Building.* MIT Press, 2008.

Pielou, E. C. *The Energy of Nature*. University of Chicago Press, 2001.

Pietsch, Theodore W. *Trees of Life: Visual History of Evolution*. Johns Hopkins University Press, 2012.

Pollan, Michael. *The Omnivore's Dilemma: A Natural History of Four Meals*. Penguin, 2006.

Pollan, Michael. *In Defense of Food: An Eater's Manifesto*. Penguin, 2008.

Rapport, D. J., Connie L. Gaudet, R. Costanza, P. R. Epstein, and R. Levins, eds., *Ecosystem Health: Principles and Practice*. Wiley-Blackwell, 1998.

Redfern, Ron. *Origins: The Evolution of Continents, Oceans and Life*. University of Oklahoma Press, 2001.

Restany, Pierre. *Hundertwasser: The Painter-King with the Five Skins*. Taschen, 2003.

Robinson, Kim Stanley. *2312*. Hachette, 2012.

Rothenberg, David. *Survival of the Beautiful: Art, Science, and Evolution*. Bloomsbury, 2011.

Ryan, Jon C., and Alan Thein Durling, *Stuff: The Secret Lives of Everyday Things*. Northwest Environment Watch, 1997.

Sagarin, Rafe, and Anibal Pauchard. *Observational Ecology: Broadening the Scope of Science to Understand a Complex World*. Island, 2012.

Sanderson, Eric W. *Manahatta: A Natural History of New York City*. Abrams, 2009.

Schafer, R. Murray. *The Soundscape: Our Sonic Environment and the Tuning of the World*. Destiny, 1994.

Schellenberger, Michael, and Ted Nordhaus, eds. *Love Your Monsters: Postenvironmentalism and the Anthropocene*. Breakthrough Institute, 2011.

Schellnhuber, Hans-Joachim, Paul J. Crutzen, William C. Clark, Martin Claussen, and Hermann Held. *Earth Systems Analysis for Sustainability*. MIT Press, 2004.

Schneider, Stephen H. *Global Warming: Are We Entering the Greenhouse Century?* Sierra Club Books, 1989.

Schultz, Kathryn. *Being Wrong: Adventures in the Margins of Error*. HarperCollins, 2010.

Senge, Peter, Bryan Smith, Nina Kruschwitz, Joe Laur, and Sara Schley. *The Necessary Revolution: How Individuals and Organizations Are Working Together to Create a Sustainable World*. Doubleday, 2008.

Serrell, Beverly. *Exhibit Labels: An Interpretive Approach*. Altamira, 1996.

Shedroff, Nathan. *Design Is the Problem: The Future of Design Must Be Sustainable*. Rosenfeld, 2009.

Shi, David E. *The Simple Life: Plain Living and High Thinking in American Culture*. Oxford University Press, 1985.

Shirky, Clay. *Cognitive Surplus: How Technology Makes Consumers into Collaborators*. Penguin, 2010.

Snyder, Gary. *The Practice of the Wild*. North Point, 1990.

Snyder, Gary. *The Gary Snyder Reader: Prose, Poetry and Translations*. Counterpoint, 1999.

Sobel, David. *Childhood and Nature: Design Principles for Educators*. Stenhouse, 2008.

Souder, William. *On a Farther Shore: The Life and Legacy of Rachel Carson*. Crown, 2012.

Steffen, W., A. Sanderson, P. D. Tyson, J. Jager, P. A. Matson, B. Moore III, F. Oldfield, et al. *Global Change and the Earth System: A Planet Under Pressure*. Springer, 2003.

Tatum, Beverly. *Why Are All the Black Kids Sitting Together in the Cafeteria?* Basic Books, 2003.

Thomashow, Mitchell. *Ecological Identity: Becoming a Reflective Environmentalist*. MIT Press, 1995.

Thomashow, Mitchell. *Bringing the Biosphere Home: Learning to Perceive Global Environmental Change*. MIT Press, 2001.

Thomashow, Mitchell. "The Gaian Generation: Toward Pattern-Based Environmental Learning." In *Gaia in Turmoil*, ed. Eileen Crist and H. Bruce Rinker. MIT Press, 2010.

Torres, Vasti, Susan R. Jones, and Kristen A. Renn. "Identity Development Theories in Student Affairs: Origins, Current Status, and New Approaches." *Journal of College Student Development* 50 (2009), no. 6: 577–596.

Volk, Tyler. *CO_2 Rising: The World's Greatest Environmental Challenge*. MIT Press, 2008.

Wackernagel, Mathis, William Rees, and Phil Testemale. *Our Ecological Footprint: Reducing Human Impact on the Earth. Gabriola Usland*. New Society, 1998.

Walker, Brian, and David Salt. *Resilience Thinking: Sustaining Ecosystems and People in a Changing World*. Island, 2006.

Weintraub, Linda. *To Life! Eco Art in Pursuit if a Sustainable Planet*. University of California Press, 2012.

Yergin, Daniel. *The Prize: The Epic Quest for Oil, Money & Power*. Simon and Schuster, 1991

Yergin, Daniel. *The Quest: Energy, Security and the Remaking of the Modern World*. Penguin, 2011.

Index